MAKING SENSE OF MATHEMATICS FOR TEACHING

Girls in Grades K–5

THOMASENIA LOTT ADAMS
TAYLAR B. WENZEL
KRISTOPHER J. CHILDS
SAMANTHA R. NEFF

Copyright © 2019 by Solution Tree Press

Materials appearing here are copyrighted. With one exception, all rights are reserved. Readers may reproduce only those pages marked "Reproducible." Otherwise, no part of this book may be reproduced or transmitted in any form or by any means (electronic, photocopying, recording, or otherwise) without prior written permission of the publisher.

555 North Morton Street
Bloomington, IN 47404
800.733.6786 (toll free) / 812.336.7700
FAX: 812.336.7790

email: info@SolutionTree.com
SolutionTree.com

Visit **go.SolutionTree.com/mathematics** to download the free reproducibles in this book.

Printed in the United States of America

Library of Congress Cataloging-in-Publication Data

Names: Adams, Thomasenia Lott, author. | Wenzel, Taylar (Taylar B.), 1981- author. | Childs, Kristopher (Kristopher J.), 1982- author. | Neff, Samantha (Samantha R.), author.

Title: Making sense of mathematics for teaching girls in grades K-5 / Thomasenia Lott Adams, Taylar B. Wenzel, Kristopher J. Childs, and Samantha R. Neff.

Description: Bloomington, IN : Solution Tree Press, [2019] | Includes bibliographical references and index.

Identifiers: LCCN 2018042847 | ISBN 9781945349782 (perfect bound)

Subjects: LCSH: Mathematics--Study and teaching (Elementary)--United States. | Arithmetic--Study and teaching (Elementary)--United States. | Girls--Education--United States. | Sex differences in education--United States. | Mathematical ability--Sex differences--United States. | Academic achievement--United States.

Classification: LCC QA135.6 .A38 2019 | DDC 372.7/2--dc23 LC record available at https://lccn.loc.gov/2018042847

Solution Tree

Jeffrey C. Jones, CEO
Edmund M. Ackerman, President

Solution Tree Press

President and Publisher: Douglas M. Rife
Associate Publisher: Sarah Payne-Mills
Art Director: Rian Anderson
Managing Production Editor: Kendra Slayton
Senior Editor: Amy Rubenstein
Developmental Editors: Ashante Thomas and Miranda Addonizio
Proofreader: Sarah Ludwig
Cover Designer: Jill Resh
Text Designer: Abigail Bowen
Text Compositor: Jill Resh
Editorial Assistant: Sarah Ludwig

We dedicate this book to each and every girl studying mathematics in classrooms around the world—and to the teachers who empower these girls.

Acknowledgments

As I mature, I realize that my journey on this planet is limited and meant to be immensely enjoyed. What gives me the most joy is family: my husband, Larry; our sons, Blake, Philip, and Kurt; and our children of love, Georges and Diana. I must also express gratefulness for my father (in memory), my mother, and my seven siblings and their families. I am also thankful to loyal and trusted friends, mentors and mentees, an awesome church family, and wonderful coauthors. I truly appreciate Juli K. Dixon and Edward C. Nolan, the *D* and *N* of Dixon Nolan Adams (DNA) Mathematics. They are cornerstones of support and give 100 percent all the time to *Making Sense of Mathematics for Teaching.*

—Thomasenia Lott Adams

The opportunity to share my passion for education would not be possible without the love, support, and sacrifices of my family. My husband, Brian, and our children, Julia, Connor, and Jake, are always at the heart of my inspiration and happiness. I am also extremely thankful for my DNA Mathematics family, who is positively impacting the experiences of teachers and learners of mathematics every day. I am honored to be a part of such a dynamic team.

—Taylar B. Wenzel

This book is dedicated to Kyndall and Kyla. You are my inspiration for all that I do. I will continue to work tirelessly to make the world a better place for you. To my lovely wife, thank you for supporting and believing in my dreams. To my mom and dad, thanks for all that you have instilled in me. To the Childs and Harris families, I love you all.

—Kristopher J. Childs

Many thanks to my husband, Jason, and our children, Ashton, Brennon, and Aubrey, for all your enthusiasm and support as I pursue my passion in engaging teachers and students in mathematics. None of this would be possible without the experiences and interactions I have had with my DNA Mathematics family, colleagues, and each student I have taught. Most importantly, Mom and Dad, you have taught me to persevere and given me the confidence to believe I can do anything. I am blessed to have all of you in my life; thank you.

—Samantha R. Neff

Solution Tree Press would like to thank the following reviewers:

Elizabeth Gehron
Mathematics Interventionist
Idyllwilde Elementary
Sanford, Florida

Vernita Glenn-White
Assistant Professor
Stetson University
DeLand, Florida

Ashauna Lindo
Virtual School Facilitator—Mathematics
Aldine Virtual School
Houston, Texas

Shauna Paedae
Mathematics Teacher on Special Assignment
Escambia County Public Schools
Pensacola, Florida

Stephanie Reddick
K–5 Mathematics Coordinator
Atlanta Public Schools
Atlanta, Georgia

Visit **go.SolutionTree.com/mathematics** to download the free reproducibles in this book.

Table of Contents

About the Authors xi

Introduction 1
   Guiding Principles 2
   Mathematics Gender Achievement Gap 2
      Perceptions 3
      Possibilities 3
      Priorities 3
   Audience 3
   What This Book Offers 4

CHAPTER 1
Mathematics Gender Achievement Gap 7
   Exploring the Mathematics Gender Achievement Gap 10
      Evidence Pointing to a Gender Gap in Mathematics 10
      Evidence Challenging a Gender Gap in Mathematics 14
   Considering the Impact of Teachers' Mindsets 16
   Conclusion 16

CHAPTER 2
Perceptions About Girls in Mathematics 19
   Tying Belief, Bias, and Stereotype to Perception 19
   Unearthing Perceptions About Girls in Mathematics 22
      The Classroom 22
      The School and District 25
      The School-Home Connection 27
   Conclusion 30

CHAPTER 3
Possibilities for Girls in Mathematics 33
   Honoring Diverse Ways of Doing Mathematics 36
      What Actions Does the Teacher Take to Engage the Girls in Learning Mathematics? 36

      What Is the Evidence That the Girls Are Actively Engaged in Learning Mathematics? 37
      How Does the Teacher Challenge the Girls to Think Critically About the Mathematics? 37
   Fostering Classroom Discourse. 37
      What Actions Does the Teacher Take to Engage the Girls in Learning Mathematics?. 38
      What Is the Evidence That the Girls Are Actively Engaged in Learning Mathematics? 39
      How Does the Teacher Challenge the Girls to Think Critically About the Mathematics? 39
   Planning for Hands-On Learning 39
      What Actions Does the Teacher Take to Engage the Girls in Learning Mathematics?. 40
      What Is the Evidence That the Girls Are Actively Engaged in Learning Mathematics? 40
      How Does the Teacher Challenge the Girls to Think Critically About the Mathematics? 41
   Using Questioning to Boost Understanding 41
      What Actions Does the Teacher Take to Engage the Girls in Learning Mathematics?. 42
      What Is the Evidence That the Girls Are Actively Engaged in Learning Mathematics? 43
      How Does the Teacher Challenge the Girls to Think Critically About the Mathematics? 43
   Using Formative Assessment 43
      What Actions Does the Teacher Take to Engage the Girls in Learning Mathematics?. 44
      What Is the Evidence That the Girls Are Actively Engaged in Learning Mathematics? 44
      How Does the Teacher Challenge the Girls to Think Critically About the Mathematics? 45
   Considering Context for Tasks 46
      What Actions Does the Teacher Take to Engage the Girls in Learning Mathematics?. 47
      What Is the Evidence That the Girls Are Actively Engaged in Learning Mathematics? 47
      How Does the Teacher Challenge the Girls to Think Critically About the Mathematics? 47
   Modeling of Mathematical Power 48
      What Actions Does the Teacher Take to Engage the Girls in Learning Mathematics?. 48
      What Is the Evidence That the Girls Are Actively Engaged in Learning Mathematics? 49
      How Does the Teacher Challenge the Girls to Think Critically About the Mathematics? 49
   Conveying Teacher Expectations 49
      What Actions Does the Teacher Take to Engage the Girls in Learning Mathematics?. 50
      What Is the Evidence That the Girls Are Actively Engaged in Learning Mathematics? 51
      How Does the Teacher Challenge the Girls to Think Critically About the Mathematics? 51
   Planning for Positive Practices for Girls Studying Mathematics 52
      The Classroom 52
      The School and District. 60
      The School-Home Connection 62
   Conclusion. 64

CHAPTER 4
Priorities for Teaching Girls Mathematics **67**
   Equity 68
      Making an Impact 70
      Using the TQE Process as a Guide for Inclusiveness of All Students 71

Teacher Beliefs 73
Opportunity 76
Teacher Knowledge. 79
Conclusion 81

EPILOGUE
Encouragement for Girls in Mathematics. 85
Resources to Encourage Girls in Mathematics 85
Conclusion. 89

References and Resources 91

Index. 101

About the Authors

Thomasenia Lott Adams, PhD, is an associate dean and professor of mathematics education in the College of Education at the University of Florida (UF). She has mentored many future teachers of mathematics and mathematics teacher educators, and has served as a mathematics coach for grades K–12. She is the author of an elementary mathematics text series, academic books, and numerous peer-reviewed journal articles. She is a featured speaker in a variety of venues.

Dr. Adams is an associate editor for the National Council of Teachers of Mathematics' (NCTM) journal, *Mathematics Teacher: Learning and Teaching PreK–12*. She previously served as editor for NCTM's Mathematical Roots Department in *Mathematics Teaching in the Middle School* and coeditor for the Investigations Department of *Teaching Children Mathematics*. She was the Program Chair for the 2018 Annual Conference of NCTM. Her professional service also includes the roles of past president of the Florida Association of Mathematics Teacher Educators, past board member for the Association of Mathematics Teacher Educators, and past board member of the School Science and Mathematics Association. She is also a past recipient of the Mary L. Collins Teacher Educator of the Year Award from the Florida Association of Teacher Educators. Dr. Adams has engaged in many other high-impact mathematics education projects. She is a leader in Dixon Nolan Adams (DNA) Mathematics and the mathematics program officer for the University of Florida Lastinger Center for Learning.

Dr. Adams received a bachelor of science in mathematics from South Carolina State College and a master of education and doctorate of philosophy in instruction and curriculum with an emphasis in mathematics education from the University of Florida.

To learn more about Dr. Adams's work, visit www.dnamath.com and follow @TLAMath on Twitter.

Taylar B. Wenzel, EdD, is a faculty member in the College of Community Innovation and Education at the University of Central Florida, where she teaches undergraduate and graduate students. Her research focuses on the role of children's use of cognitive and metacognitive strategies across reading and mathematics contexts and the use of lesson study with preservice teachers.

Dr. Wenzel is well known for her professional development partnerships with urban schools and programs in the Orlando, Florida, area. At each site, she teaches model lessons, facilitates professional learning, conducts research, and supports instructional change. She is the cofounder of the UCF Enrichment Programs in Literacy and Mathematics, a collaboration with other faculty members, through which undergraduate and graduate students provide intervention to students in the Central Florida area.

Dr. Wenzel regularly presents her work at professional conferences and invited keynote addresses. She is a consultant for numerous school districts and education agencies and has published articles, book chapters, and professional development handbooks.

Dr. Wenzel earned bachelor's and master's degrees in elementary education with an emphasis in mathematics and science education from the University of Florida and a doctorate in curriculum and instruction from the University of Central Florida.

To learn more about Dr. Wenzel's work, follow @taylar_wenzel on Twitter.

Kristopher J. Childs, PhD, is a mathematics educator focused on inspiring change through mathematics. His work centers on helping educators understand and teach mathematics effectively. Dr. Childs is actively researching the selection, implementation, and discourse of rich problem-solving tasks and teaching mathematics for social justice.

He has vast experience at the K–12 and postsecondary levels in teaching and leadership positions. His experiences have afforded him the opportunity to gain hands-on, practical application in a variety of educational settings working with diverse student populations. He is committed to ensuring every student receives a high-quality mathematics education.

Dr. Childs earned a bachelor of science degree in computer engineering from Florida Agricultural and Mechanical University. He completed a master of science degree in mathematics education at Nova Southeastern University and received a doctorate of philosophy in mathematics education from the University of Central Florida.

To learn more about Dr. Childs's work, visit www.KristopherChilds.com and follow @DrKChilds on Twitter.

Samantha R. Neff is an adjunct professor in early childhood education at the University of Central Florida. She has been a collaborative team member in Seminole County Public Schools since 1997, writing mathematics instructional plans and participating in grant studies on formative assessment in mathematics and cognitively guided instruction. She is an elementary mathematics instructional coach in a Title I school. Neff works with teachers, students, parents, and administrators to enhance the teaching, learning, and assessing of mathematics to improve student achievement. Focused on mathematical practices that promote critical thinking and reasoning in the classroom, she has organized and implemented a flourishing professional learning community. Prior to coaching, Neff worked in K–5 classrooms for more than ten years. She is also a consultant for specializing in building mathematical content knowledge and assessment.

Neff was selected as 2019 District Teacher of the Year in Seminole County, Florida, and was a finalist for 2019 Florida Teacher of the Year. She earned bachelor's and master's degrees in early childhood education from the University of Central Florida.

To learn more about Neff's work, follow @jasamneff on Twitter.

To book Thomasenia Lott Adams, Taylar B. Wenzel, Kristopher J. Childs, or Samantha R. Neff for professional development, contact pd@SolutionTree.com.

Introduction

Juli K. Dixon, Edward C. Nolan, and Thomasenia Lott Adams conceptualized the *Making Sense of Mathematics for Teaching* books as a "response to requests from teachers, coaches, supervisors, and administrators who understand the need to know mathematics for teaching but who do not know how to reach a deeper level of content knowledge or support others to do so" (Dixon, Nolan, Adams, Brooks, & Howse, 2016, p. 1). These books provide "guidance for refining what it means to be a teacher of mathematics. To teach mathematics for depth means to facilitate instruction that empowers students to develop a deep understanding of mathematics" (Dixon, Nolan, Adams, Brooks, & Howse, 2016, pp. 1–2).

The positive reaction to this series has strengthened our commitment to this message and has fueled our goal of improving the teaching and learning of mathematics worldwide.

Every endeavor has its teachable moments, and what we have learned since the first *Making Sense of Mathematics for Teaching* book's publication is that teachers, coaches, supervisors, and administrators also desire to learn more about teaching and learning mathematics to distinct populations of students, such as students with special needs and English learners. In fact, *Making Sense of Mathematics for Teaching the Small Group* (Dixon, Brooks, & Carli, 2019) is an example of our responsiveness to the specific interests and needs of teachers who desire information about effective practices for teaching mathematics to students in the pulled small group. The context for this book, *Making Sense of Mathematics for Teaching Girls in Grades K–5*, is also a response to those who called for focus on a distinct population of students of mathematics: girls.

Comparing boys and girls is a common phenomenon in various contexts in life, and so it is not surprising that questions about gender differences are present in discourse about schooling. We acknowledge that gender differences as presented in the research may arise socially, culturally, and in other experiences that interplay with gender identity. The interest in girls in the mathematics classroom emerges from long-term and ongoing results of education research and education testing that use gender as a variable to study the impact of classroom instruction on students' learning and achievement (among other things) in mathematics. Questions that research and testing often ask are designed to describe possible differences in mathematics learning and achievement on the variable of gender to determine what leads to these differences, and to find ways to close any gaps on achievement present for the variable of gender. We in no way intend to discredit, dismiss, or disqualify any children and how they may identify regarding gender. We in no way attempt to address the complexities that might arise from gender identification. We definitely do not intend any offense toward any person. We offer our thoughts and ideas for the sole purpose of supporting all students to learn and succeed in mathematics.

Our general takeaway from countless discussions with colleagues in the field is that girls sometimes have less interest and focus in mathematics as they progress in school, especially when moving to middle and high school. We see assessment data at local levels concluding that girls' achievement in mathematics is not what is desired and is not on the same level of achievement as boys. Colleen M. Ganley and Sarah Theule Lubienski (2016b) shed light on the importance of this matter:

> Given that achievement is a consistent predictor of girls' later interest and confidence in math, even after conditioning on current interest and confidence, [their] study suggests that small gender differences in early achievement could exacerbate later differences in interest and confidence. Thus, increasing girls' achievement is critical for later achievement and math attitudes, and early math confidence and interest are also important. (p. 190)

This book is a response to teachers, coaches, supervisors, and administrators who seek support for addressing the needs of girls in the mathematics classroom in grades K–5. Our intent is to confront the challenge at the elementary level to serve as a stopgap for issues in the later grades and beyond.

Guiding Principles

We used several guiding principles to write this book. First, in no way do we support separating mathematics content by gender. Good mathematics is good mathematics. Good teaching of mathematics is good teaching of mathematics. A learner of mathematics is a learner of mathematics. There is not a body of mathematics for boys and a different set of mathematics for girls. Mathematics, in its entirety, is for everyone, and we believe that a deep understanding of mathematics is a valuable asset. If you desire an opportunity to make sense of *mathematics* for teaching, we refer you to the *Making Sense of Mathematics for Teaching* grade-band books (Dixon, Nolan, Adams, Brooks, & Howse, 2016; Dixon, Nolan, Adams, Tobias, & Barmoha, 2016; Nolan, Dixon, Roy, & Andreasen, 2016; Nolan, Dixon, Safi, & Haciomeroglu, 2016) of interest, in which the authors address mathematics content and mathematical pedagogy in great detail. Our aim in *Making Sense of Mathematics for Teaching Girls in Grades K–5* is to follow up the series with a spotlight on particular perspectives and instructional moves that give space to girls to support their learning and achievement of mathematics.

Secondly, in the course of applying the ideas we present, we encourage you to include *all* students in your classroom. In no way do we want to say that our ideas are for girls only! By all means, apply any ideas from this book with any students you teach. Our greatest reward is for all students to succeed in mathematics. However, in sharing this book and its focus with you, we are only asking that you take the time to think about the ways you can make engaging girls in mathematics a better and more positive experience. We want girls to have an opportunity to overcome challenges to learning mathematics that are often present when gender (among other considerations) is a factor.

Finally, while we want to inform you about the mathematics gender achievement gap data on the variable of gender, we also want to position you to think deeply about your own and others' perceptions about girls in mathematics. We follow this course because so much of what transpires for girls in mathematics is not data-driven but perception-driven. We hope this book will be a catalyst for how we think about girls in mathematics, because how we think influences how we behave and how we interact with others.

Mathematics Gender Achievement Gap

The foundation for our discussion is the gender achievement gap because this is a construct that most often drives discourse about girls and boys in mathematics. Here are two important questions: (1) Does there presently exist a mathematics achievement gap between girls and boys? and (2) If so, which gender has the highest achievement? Well, it depends! It depends on other variables (for instance, race, socioeconomics, and so on) and the subsequent data that are under consideration. We will revisit this topic more carefully in chapter 1.

We approach the focus on the mathematics gender achievement gap by using three lenses: (1) perceptions, (2) possibilities, and (3) priorities. We now offer clarity on these terms and how we will use them throughout the book.

Perceptions

Perceptions refer to beliefs, biases, and stereotypes that individuals hold. You will read about perceptions in terms of how girls perceive themselves as learners of mathematics, and you will also read about the perceptions of others, such as teachers. We encourage you to consider the important role that perceptions play in impacting girls and their relationship with mathematics. Our position is that perceptions play a key role in girls' experiences in mathematics because teachers' and girls' attitudes, perspectives, and thoughts about and toward one another are present during these interactions.

Possibilities

We use the term *possibilities* to describe actionable activities and behaviors that educators can implement to have a potentially positive impact on girls as learners of mathematics. For example, possibilities in classrooms and schools will involve instructional activities that teachers and coaches can immediately put into action with the students in their classrooms. In many cases, the possibilities this book offers also have an impact on the perceptions of and about girls in mathematics. For specific mathematics tasks and lessons, we refer you to the *Making Sense of Mathematics for Teaching* grade-band books or mathematics subject of interest. For this book, our focus is on how to ensure girls receive the attention they need to support their engagement in mathematics. Another reason we use possibilities is that we do not want to limit you in the strategies and mindsets you might apply to support girls in mathematics. We realize there may be many factors that you must consider given your local context. We simply ask that you join us in thinking about and doing what's possible to help girls succeed in mathematics.

Priorities

We propose that there are *priorities* around such matters as structures, policies, and systems that impact the access to and quality of mathematical learning experiences for girls. We suggest several priorities that should be present in decision making regarding teaching and learning mathematics that influence the mathematics experiences of girls. These priorities are intended to support classroom experiences that empower students to learn mathematics. Priorities could be relevant to teachers, coaches, school leaders, and district leaders as they make decisions that impact students' learning of mathematics. It is our hope that the discussion of priorities will have an impact on what's possible and the perceptions that relate to mathematics experiences for girls.

Audience

We wrote this book particularly for teachers, coaches, supervisors, and administrators who regularly influence the mathematics learning experiences of girls in grades K–5, either directly or indirectly. If you are a teacher who is preparing to teach or currently teaching mathematics, we hope that you will read this book with your students in mind and look for opportunities to expand your own teaching practices to strengthen the mathematics experiences of all your students, but specifically girls. If you are in a coaching, supervising, or administrative role, we hope that you will read this book with your teachers in mind as you consider how you can support them and positively impact instructional change in mathematics. Consider the ways that you might use your role and strategies you learn from this book to directly improve mathematics learning experiences for all students, and particularly girls.

What This Book Offers

We organized the book so that the information we share helps you build knowledge that is helpful for supporting girls in mathematics. Its four chapters answer four questions, respectively.

1. What do educators know (or think they know) about girls learning mathematics?
2. Why do *perceptions* about mathematics and girls learning mathematics matter—whether these are perceptions teachers, girls, or others hold?
3. What *possibilities* of actions and activities might promote girls' success in mathematics?
4. How might *priorities* support and strengthen girls' experiences as learners of mathematics?

Chapter 1 digs into what educators, particularly mathematics teachers, coaches, and administrators, know (or think they know) about mathematics teaching and learning practices for girls. It sets the stage, providing a foundation on the mathematics gender achievement gap and highlighting the contributors to this gap. Chapters 2 and 3 focus on perceptions and possibilities; specifically, chapter 2 explores the question of *why* teachers', girls', and other parties' perceptions about girls learning mathematics matter. Chapter 3 informs readers of the possibilities (actionable activities and behaviors) that might promote girls' success in mathematics. In chapters 2 and 3, we discuss issues across three contexts: (1) the classroom, (2) the school and district, and (3) the school-home connection. In chapter 2, we unpack the topic by identifying a relevant framework, norms, and impacts. Approaching the topics in this manner allows you to consider the various ways that these contexts impact girls' mathematics learning. It also allows you to focus on the parts of this book that are most relevant to you based on your role and relationship with girls in mathematics. Chapter 3 also discusses planning for positive practices for girls studying mathematics within the context of the classroom, the school and district, and the school-home connection. Chapter 3 includes videos to provide a window into classrooms where girls are active and engaged participants, where girls are challenged to learn, and where girls benefit from mathematics instruction that is designed for students to develop conceptual understanding of mathematics. While both boys and girls are present in the classroom videos, our focus is on observing the girls to learn more about how girls can engage in mathematics confidently and successfully. The *play button* symbol (figure I.1) indicates that an online video depicting mathematics instruction is available for you to watch. You can find the videos either by scanning the adjacent QR code or by following the provided URL. In chapter 4, we present a variety of priorities that will help you position yourself as an advocate for girls in mathematics. You'll gain tools to strengthen girls' experiences as learners of mathematics. The epilogue offers motivation and affirmation that you can use to further support girls in having positive experiences with and in mathematics.

Figure I.1: *Play button* symbol.

We also incorporate a variety of pedagogical features throughout the book to help clarify our ideas. For example, each chapter includes reflections from teachers, parents, or students to help you reflect on your own thinking and connect more personally with the book and to others who might be reading and discussing the book with you. There's also opportunity for you to reflect on the chapter's theme before diving into its content. We invite you to use these reflections as conversation starters with other teachers, coaches, supervisors, and administrators who engage you in conversation about girls in mathematics. Other pedagogical features include:

- **Do Now**—As you read, you will notice figures called *Do now*, which instruct you to respond to questions, prompts, or both in relation to the content. These figures are opportunities to improve your reading experience, increase your takeaways from reading, and kick-start team discussion of student experiences in mathematics. See figure I.2 for the *Do now* symbol.

Figure I.2: *Do now* symbol.

- **Take Action**—Each chapter includes figures with concrete actions that you can implement based on your role to make a difference in girls' mathematics learning experiences. Of course, you may modify these actions to better fit for your context and concern. This is a place where you can consider possibilities.

- **Reflections**—End-of-chapter reflection questions help you personally reflect on what you have read and how it aligns to your own personal perceptions, possibilities, and priorities. Use these questions to challenge yourself.

- **Further Reading**—Finally, in each chapter we recommend research and data to the extent that you will be informed but not overwhelmed. You can use these suggested articles and books to gain more insight from the field. Go online to **go.SolutionTree.com/mathematics** to find these compiled into a single reproducible document.

We developed this book to provide a lens into girls learning mathematics. As you read through it, take time to truly reflect upon the ideas we present. Consider how you will make changes to your instruction and interactions with all students in your classroom as you consider equity across gender. The key is understanding that this will be a process that will improve over time. Allow yourself the opportunity and the freedom to grow and develop as an educator. Be patient and know that over time you will transform as an educator and most importantly the girls you serve in your respective environments will have transformational mathematics experiences. To begin this journey, let's first explore the mathematics gender achievement gap.

CHAPTER 1
Mathematics Gender Achievement Gap

Some things will drop out of the public eye and will go away, but there will always be science, engineering, and technology. And there will always, always be mathematics.

—Katherine Johnson, Mathematician and Recipient of the Presidential Medal of Freedom

What do educators know (or think they know) about girls learning mathematics? This chapter will address this question through the primary lens of the much-talked-about mathematics gender achievement gap. We will also begin discussion about the influence of teachers on girls' learning mathematics and deepen our discussions on what we mean by perceptions, possibilities, and priorities.

First, however, start this journey by considering what some are saying about girls in mathematics. Perhaps you can recollect statements from conversations with friends and family, from the media, or unsolicited from strangers who learn that you are involved with the teaching and learning of mathematics. Complete the chart in figure 1.1 to record some of these recollections about girls in mathematics.

What does society (for example, media) say about girls in mathematics? • • •	What do teachers say about girls in mathematics? • • •
What do parents and guardians say about girls in mathematics? • • •	What do girls say about girls in mathematics? • • •

Figure 1.1: Educator recollections of what's said about girls in mathematics.

*Visit **go.SolutionTree.com/mathematics** for a free reproducible version of this figure.*

Circle the statements in figure 1.1 that are positive. How many of the statements did you circle? We would like all the statements to be positive, but that is most likely not the case. Often this is due to the negative portrayal of girls in mathematics. Many times girls have seen the field of mathematics portrayed as a male-dominated field, thus they are typically excluded from the narrative. Nonetheless, what we strive for is an environment that supports girls studying mathematics and for girls to receive positive influences to continue their trajectory in mathematics. When what is said about girls in mathematics is positive, the actions taken to support girls are more likely to be positive. Furthermore, because what people say to and about children easily influences them, the more we can make sure girls hear positive statements about their learning of mathematics, the better chance we have of positively influencing girls in their study of mathematics. (Who else would you consider for what they have to say about girls in mathematics? Feel free to replicate figure 1.1, page 7, using other groups of people to further the discussion.)

Before we continue, we ask you to share your own feelings (figure 1.2). Reflect on your own thoughts about girls as learners of mathematics using the questions and prompts. Your responses will help you examine and frame your thinking and reflections as you continue to read this book and as you teach mathematics to girls. If you are engaged in a book study with *Making Sense of Mathematics for Teaching Girls in Grades K–5*, you can also use the items to promote helpful discussions about girls' experiences with mathematics.

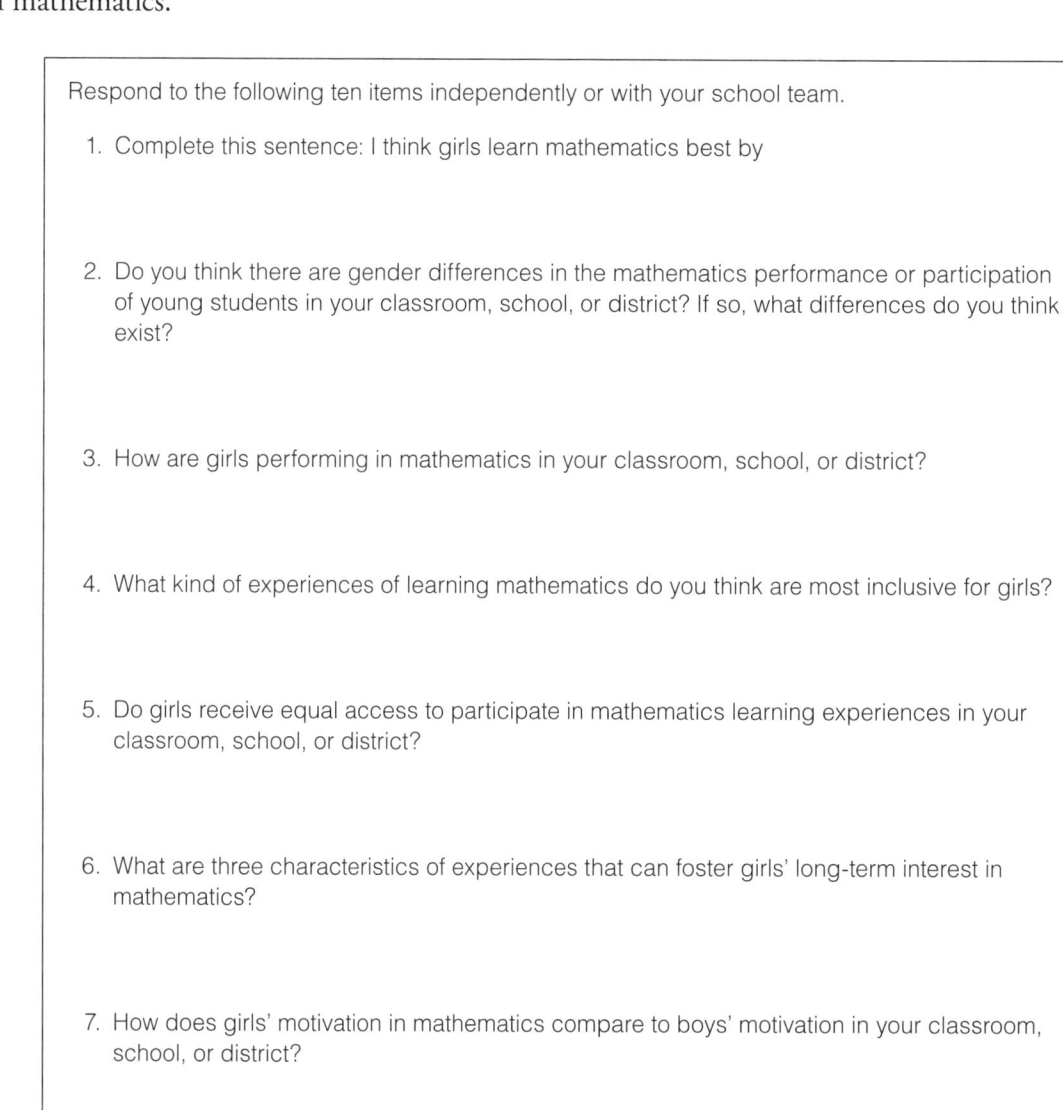

Respond to the following ten items independently or with your school team.

1. Complete this sentence: I think girls learn mathematics best by

2. Do you think there are gender differences in the mathematics performance or participation of young students in your classroom, school, or district? If so, what differences do you think exist?

3. How are girls performing in mathematics in your classroom, school, or district?

4. What kind of experiences of learning mathematics do you think are most inclusive for girls?

5. Do girls receive equal access to participate in mathematics learning experiences in your classroom, school, or district?

6. What are three characteristics of experiences that can foster girls' long-term interest in mathematics?

7. How does girls' motivation in mathematics compare to boys' motivation in your classroom, school, or district?

8. Why do you think some people so easily say things like, "I'm just not good at mathematics"?

9. Do you think teachers' perceptions of mathematics impact girls' perceptions of mathematics? If so, why and how?

10. What positive statement or action have you recently said or done to encourage a girl to succeed in mathematics?

Figure 1.2: Educator reflection on the gender gap in mathematics learning.

*Visit **go.SolutionTree.com/mathematics** for a free reproducible version of this figure.*

Consider the voices of two elementary school girls as they reflect on their experiences with mathematics. Christina, a kindergarten student, says:

> When I do math in school, I write numbers, and I can say them and count them and stomp them and clap them, and learning math is fun. . . . I think I will use math when I'm older to make money and be able to go places, and I will know how to do math to show other teachers all the math I know. (C. Latanza, personal communication, September 5, 2017)

Now, consider fifth grader Julia's remarks:

> At school when I learn math, I think that it's fun, especially when I get to pick the strategy that works best for me to do the math. . . . I don't really know all of the math that I would need yet (when I grow up), but I think I might want to be an engineer, and I know that I would have to use math every day for the planning and the building that I would be doing in that job. (J. Clements, personal communication, December 2, 2017)

Use figure 1.3 to detail what you would say to these students to maintain their interest.

 What responses might you offer these young girls to support them and continue their interest in learning mathematics?

Figure 1.3: Remarks to maintain girls' interest in mathematics.

As you read the girls' perspectives of learning mathematics, you heard ideas of positivity and promise that these girls will continue to pursue opportunities for mathematics in their schooling and beyond. However, whether or not these girls will continue to achieve and believe in themselves as mathematics learners in the pivotal years to come is unknown because it is a continuous process that will take time. The key is exposing girls to opportunities; however, ultimately it is their decision whether to pursue the opportunity. We have hope! Their early experiences are already impacting the likelihood that they will pursue coursework and careers in the field of mathematics, even in implicit ways that these girls may not yet realize. We want these girls and all other girls to realize their potential in mathematics.

Exploring the Mathematics Gender Achievement Gap

Gender differences in mathematics achievement in North America have been widely discussed and studied (Cheema & Galluzzo, 2013; Damarin & Erchick, 2010; Fryer & Levitt, 2010; Leyva, 2017; Lubienski, Robinson, Crane, & Ganley, 2013; Marks, 2008; Penner & Paret, 2008; Riegle-Crumb & Humphries, 2012; Robinson & Lubienski, 2011). However, differences in population, test formats, content assessed, and other variables yield results that are not necessarily generalizable and at times even offer mixed results, thus complicating the discussion and making implications fuzzy. In the sections that follow, we will explore both sides of the issue—data that say there *is* a mathematics gender achievement gap and data that contend there *is not* a mathematics gender achievement gap. While the primary focus of this book is on girls in grades K–5, it is important to understand the broader discussion on the mathematics gender achievement gap in the context of gender differences that appear in college and career settings.

Evidence Pointing to a Gender Gap in Mathematics

In this section, we'll dive deeper into the research that shows there *is* a mathematics gender achievement gap. We'll explore the representation of women in science, technology, engineering, and mathematics (STEM) fields; differences in mathematics achievement scores; differences in student responses regarding self-concept in mathematics; differences in problem-solving approaches among boys and girls; and differences in spatial skills among boys and girls.

Representation of Women in Science, Technology, Engineering, and Mathematics Fields

The representation of women in college programs and career pathways related to science, technology, engineering, and mathematics is societal evidence of a mathematics gender gap. Catherine Riegle-Crumb and Barbara King (2010) and many other researchers suggest that there is a disproportionally low number of women (compared to men) in both STEM programs in colleges and universities and in STEM careers (Lubienski et al., 2013; Mendick, 2005; Riegle-Crumb & Humphries, 2012; Snyder & Dillow, 2011).

In regard to STEM college programs, Ryan Noonan (2017) reports that "while nearly as many women hold undergraduate degrees as men overall, they [women] make up only about 30 percent of all STEM degree holders" (p. 1). In order for any student to pursue a STEM degree in college, he or she benefits from having a good school background in STEM subjects, mathematics being one such subject.

In this same report, "Women in STEM: 2017 Update," Noonan (2017) also addresses the presence of women in STEM careers:

> Women filled 47 percent of all US jobs in 2015 but held only 24 percent of STEM jobs. Likewise, women constitute slightly more than half of college educated workers but make up only 25 percent of college educated STEM workers. (p. 1)

Hence, the gender imbalance among STEM degrees is reflected in the gender imbalance in STEM careers. Every opportunity to encourage girls to have interest in and study fields in STEM, specifically mathematics, provides an opportunity to increase the representation of women in STEM. As it relates to representation, the quote by Marian Wright Edelman (2015), "It's hard to be what you can't see," comes to mind. Therefore, it is important that girls have positive role models in their respective fields who they can look up to and follow.

You may ask why this is happening. Why are there more men than women studying in STEM programs, such as mathematics? Why are there more men than women employed in STEM fields? Economist and statistician David Beede and colleagues (2011) suggest that "there are many possible factors contributing to the discrepancy of women and men in STEM jobs, including: a lack of female role models, gender stereotyping, and less family-friendly flexibility in the STEM fields" (p. 1). While the key factor (or possible intersection of multiple factors) cannot be confirmed, we believe that continuing the dialogue and inquiry around the representation of women in STEM is warranted.

Use figure 1.4 to brainstorm other factors that influence the number of women in STEM fields.

What other factors might impact the number of girls who go on to study in STEM fields or gain employment in STEM careers?

Figure 1.4: Reflections on the factors that impact the number of women in STEM.

Differences in Mathematics Achievement Scores

The National Assessment of Educational Progress (NAEP), which assesses students in grades 4, 8, and 12, is the most commonly cited U.S. assessment in mathematics. NAEP has been administered approximately every four years since 1973; however, there have been changes in test administration since its inception that impact statistical comparisons. For example, since 1990, students have been assessed by grade rather than by age, and there was variation in the allowance of accommodations at some testing sites in 1996 and 2000.

From 2003 to 2017, there has been a slight gender difference based on the average fourth-grade mathematics assessment scores, with boys scoring significantly higher than girls during each assessment administration (National Center for Education Statistics [NCES], 2017). For example, in 2003, boys scored higher (with statistical significance) when scores were analyzed by average scale score, percentile, and proficiency level. There were significant differences that favored the performance of boys at the 25th, 50th, 75th, and 90th percentiles, and the gap between boys' and girls' scores increased as the scores increased (with a five-point difference at the 90th percentile). According to the NAEP proficiency data from 2003, boys also outperformed girls in the categories of advanced (5 percent compared to 3 percent), at or above proficient (35 percent compared to 30 percent), and at or above basic (78 percent compared to 76 percent; NCES, 2017). For example, the average scale score of fourth-grade boys was 236 compared to the average scale score of 233 for fourth-grade girls, a seemingly small but statistically significant difference. Additionally, according to the proficiency data from 2003, fourth-grade boys outperformed girls in four of the five content strands: number and operations, data analysis, algebra and functions, and measurement, with the discrepancy in measurement being the largest (NCES, 2017). This is consistent with gaps in the measurement strand from the 1996 administration of the NAEP, which Ellen Ansell and Helen M. Doerr (2000) analyze to reveal that fourth-grade boys were more accurately able to choose appropriate units or read and use a measuring instrument (such as a speedometer, thermometer, or ruler). When Ansell and Doerr (2000) further analyze the gender gaps within racial groups and across content strands, they find significant differences that still favored boys for white and Hispanic groups in number operations and measurement and Asian and Pacific islanders in measurement. This analysis also reveals, however, that African American girls outperformed African American boys in geometry and data analysis at the fourth-grade level.

While NCES (2017) documents that fourth-grade boys achieve higher scores in mathematics than girls, the achievement gap between boys and girls has not widened between 2003 and 2017, with the most recent 2017 data revealing an average scale score of 241 among boys and 239 among girls. This means that the mathematics achievement gap between grade 4 boys and grade 4 girls persists, but it has not grown in recent years.

There is more research that speaks to the matter of the mathematics gender achievement gap. For instance, Sean F. Reardon, Erin M. Fahle, Demetra Kalogrides, Anne Podolsky, and Rosalía C. Zárate (2018) report on a study ("Gender Achievement Gaps in U.S. School Districts") of students in grades 3–8 across ten thousand U.S. school districts:

> Both math and ELA gender achievement gaps vary among school districts and are positively correlated—some districts have more male-favoring gaps and some more female-favoring gaps. We find that math gaps tend to favor males more in socioeconomically advantaged school districts. (p. 2)

More specifically, among Reardon, Fahle, et al.'s (2018) findings, the distribution of mathematics gaps "implies that 95% of districts have math gaps that are between -0.07 and +0.13 standard deviations, favoring males in 72% of school districts and females in 28%" (p. 21). Additionally, this research finds that in wealthier districts and districts with more economic inequality among adult men and women, mathematics gaps favored boys on average. These analyses show that the mathematics gender achievement gap is not necessarily across the board or applicable for all students.

When considering the research presenting differences in mathematics achievement scores between boys and girls, we find that if there are differences, they are often small and are typically evident among higher-performing students (Lindberg, Hyde, Petersen, & Linn, 2010; Reardon, Fahle, et al., 2018).

Differences in Student Responses Regarding Self-Concept in Mathematics

In addition to achievement data, NAEP (NCES, 2017) reports data based on students' questionnaire responses about their beliefs about mathematics and themselves as learners. For example, when asked to consider the statement "I am good at mathematics," students could choose the answers "A lot like me," "A little like me," or "Not like me." Among fourth-grade students, boys were significantly more likely than girls to identify the following statements as being a lot like themselves: "I like mathematics" (50 percent boy, 43 percent girl), "I am good at mathematics" (56 percent boy, 43 percent girl), and "I understand most of what goes on in mathematics class" (58 percent boy, 55 percent girl).

Additional data from the Education Quality and Accountability Office (Casey, 2017) support the idea that gaps in students' self-concept may not be limited to the United States. For example, although girls and boys earned similar grades during the 2016 to 2017 academic year:

> Only 49 percent of Grade 3 girls in Ontario agreed that they were good at math compared to 62 percent of boys. The difference widens in Grade 6, where 46 percent of girls said they were good at math compared to 61 percent of boys. (Casey, 2017)

Differences in Problem-Solving Approaches Among Boys and Girls

In 1980, *Problem Solving in School Mathematics* (Krulik & Reys, 1980) initiated a shift in mathematics education that proposed problem solving to be central to mathematics instruction and across mathematics curriculum. Along with this notion, the discussion of methods, strategies, and heuristics for problem solving abound in mathematics publications and conference presentations. In addition, starting from this point, research on problem solving became more visible in the discipline. For example, Elizabeth Fennema, Thomas P. Carpenter, Victoria R. Jacobs, Megan L. Franke, and Linda W. Levi (1998) find that boys were more likely than girls to use novel or invented problem-solving approaches when given mathematics tasks. Comparison observations find that girls were more inclined to use the specific procedures that the teacher taught in previous instruction for a given problem type. The researchers further explain that the use of invented algorithms appeared to be important for students to develop key concepts in mathematics, such as place value and number sense, and for students to be flexible in new situations, such as extensions of learned mathematics. Ana Villalobos (2009) offers additional research findings that explore "strategy socialization" with regard to risk-taking and rule-following, and suggests that girls are disproportionally represented in the development of "algorithmic strategies" and boys in "problem solving strategies" (p. 27). In this study, the author suggests that over-rewarding a single strategy, especially when the strategy yields accurate solutions, can lead to difficulties in switching strategies, which is necessary when "solving unfamiliar problems that require new approaches later in the curriculum" (Villalobos, 2009, p. 27).

Although this research took place prior to specific curriculum standards that advocate for multistrategy instruction and problem-solving experiences, it suggests that additional research is warranted to determine if girls are unintentionally limiting their own explorations in problem solving in the classroom with their inclination to follow taught procedures. The different ways that boys and girls engage in problem solving may affect how they use problem solving to learn mathematics.

It is also important to consider the role of the teacher in girls' engagement with problem solving. *Education Week* reporter Sarah Schwartz (2018) challenges us to consider this:

> Students in classes where teachers have a "multi-dimensional" approach to problem solving that allows for multiple strategies are more likely to have a growth mindset at the end of the course than students of teachers who value speed or memorization. This effect can be more pronounced for some students than others. For example, separate research found that when female teachers had more anxiety around doing math, the girls in their classes had lower achievement. The boys in their classes did not see these same negative effects.

Sian L. Beilock, Elizabeth A. Gunderson, Gerardo Ramirez, and Susan C. Levine (2010) also note this correlation between a woman teacher's confidence with mathematics and her students' confidence. Given that most elementary teachers in the United States and Canada are women (Organisation for Economic Co-operation and Development [OECD], 2016), we can say with certainty that women teachers have a great reach in their access and interactions with learners. Therefore, it is important to consider this research that suggests that how a teacher responds to mathematics is an issue that can impact how certain populations of students, such as girls, will respond to mathematics.

Differences in Spatial Skills Among Boys and Girls

Some research suggests that boys demonstrate more sophisticated spatial skills than girls (Klein, Adi-Japha, & Hakak-Benizri, 2010). The National Council of Teachers of Mathematics (NCTM, 2000) calls for instructional programs from prekindergarten through twelfth grade to enable each and every student to "use visualization, spatial reasoning, and geometric reasoning to solve problems." A gender gap in this area is noteworthy because in addition to being a part of content standards at all grade levels, research shows that greater spatial skills are a predictor of higher mathematics performance in later years of schooling and that they also positively impact the selection of STEM-related careers (Tzuriel & Egozi, 2010). Despite this reported gender imbalance, much of this same research suggests targeted intervention can improve girls' deficits in spatial skills, even to the extent that it eliminates gender discrepancy. There is an impact on girls' exposure to the instructional experiences that have the potential to positively impact the development of students' spatial skills. To foster these experiences, teachers can:

- Explain to young people that spatial skills are not innate but developed.
- Encourage children and students to play with construction toys, take things apart and put them back together again, play games that involve fitting objects into different places, draw, and work with their hands.
- Use handheld models when possible (rather than computer models) to help students visualize what they see on paper in front of them. (Hill, Corbett, & St. Rose, 2010, p. 56)

Use figure 1.5 to reflect on why the gender achievement gap appears in some contexts but not others.

 Why do you think the mathematics gender achievement gap is present in some contexts but not in others?

Figure 1.5: Reflection on disparities in gender achievement gap across contexts.

Evidence Challenging a Gender Gap in Mathematics

The following is evidence that challenges the notion of a gender gap in mathematics. Specifically, we highlight marginal differences between boys' and girls' mathematics achievement scores and student confidence levels impacting mathematics achievement.

Marginal Differences Between Boys' and Girls' Mathematics Achievement Scores

As previously indicated, some data challenge the presence of a mathematics gender achievement gap. For instance, Jennifer E. V. Lloyd, John Walsh, and Manizheh Shehni Yailagh (2005) conducted an analysis of grades, standardized test scores, and self-efficacy responses among sixty-two fourth graders and ninety-nine seventh graders and conclude that "girls' mathematics achievement met or exceeded that of boys'" (p. 384).

The NCES (2017) report of NAEP data for 2003–2017 indicates that there is not a mathematics gender achievement gap except for grade 4, as we previously outlined. Although NCES (2017) reports the presence of the gap for just one grade, that is one grade too many! Why this phenomenon exists for grade 4 in particular is a point for further study. A variety of factors, such as grade 4 mathematics curriculum, assessment item structures, and much more, could influence this outcome, and researchers continue to explore it (Reardon, Kalogrides, Fahle, Podolsky, & Zarate, 2018). Nonetheless, it is clearly important for educators to take a closer look at the mathematics experiences of girls in elementary school, which is why we chose to focus this book on grades K–5.

Student Confidence Levels Impacting Mathematics Achievement

There are other variables that need examination. For instance, what we tell students about their mathematics performance or about anticipating their mathematics performance really matters. How can we encourage students? How can we empower students to be successful in mathematics, simply by what we say to them? How can we even the playing field for boys and girls in mathematics? Perhaps there are many answers to these questions and to the previous questions. Here is one simple answer: "When test administrators tell students that girls and boys are equally capable in math, . . . the difference in performance essentially disappears" (Hill et al., 2010, p. xv). So, what we say to students can have a great deal of influence on how they perform in mathematics. To send direct messages about the nature of intelligence as dynamic and to reduce stereotypes, teachers and administrators can:

- **Teach students that intellectual skills can be acquired**—Explain that, like muscles, the more we use our brains, the stronger they become. Help students learn that their brains form new connections as they stretch themselves and work hard to learn something new.

- **Praise students for their effort (not their outcomes)**—Give feedback about students' processes and how they arrive at conclusions.

- **Highlight the role of struggle in education**—Help convey to students that challenges, hard work, and mistakes are valuable and admirable. Explain to them that the process of struggling and overcoming challenges has been at the core of most scientific and mathematical contributions in our society.

Use figure 1.6 to rate your confidence teaching and learning mathematics.

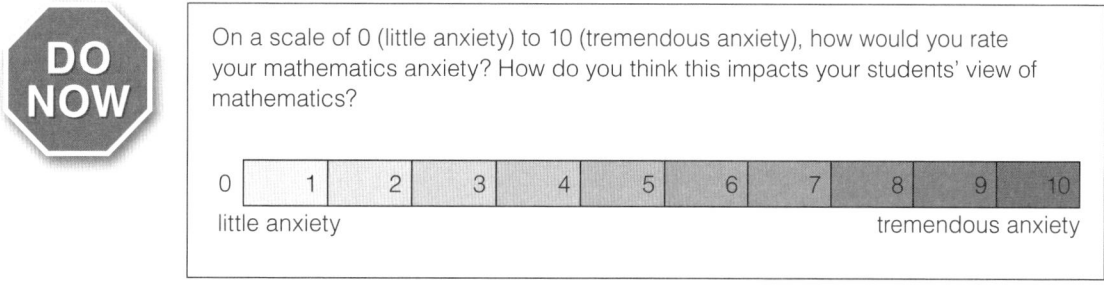

Figure 1.6: Educator reflection on anxiety toward mathematics and its impact on students.

Considering the Impact of Teachers' Mindsets

We briefly mentioned the teacher's influence in the previous section. Here we deal with this topic in a bit more detail. Elementary mathematics teachers play an important role in the mathematics learning experiences of young girls. Interestingly, studies even suggest that teachers may impact girls' perceptions of and achievements in mathematics beyond the mathematics lessons taught in the classroom, especially if that teacher is a woman (Beilock et al., 2010; Klass, 2017). Regardless of teacher effectiveness, girls' mathematics achievement may actually be lower in classrooms where a woman teacher has mathematics anxiety, meaning that she is not confident in either her own mathematics abilities, her ability to teach mathematics, or both (Beilock et al., 2010). Young girls may implicitly be forming a gender stereotype since they assume the teacher's knowledge and ability to learn applies to them.

Teachers' perceptions of student achievement in mathematics also offer potential for gender gaps. Joseph P. Robinson-Cimpian of Economics and Education Policy at New York University Steinhardt and colleagues Sarah Theule Lubienski, Colleen M. Ganley, and Yasemin Coper-Gencturk (2014) explore how teachers engage in unintentional "differential ratings," comparing teachers' projections of their students' mathematics achievement scores to their actual scores (p. 1264). In their findings (Robinson-Cimpian et al., 2014), teachers perceived boys' mathematical performance to be higher than their girl counterparts, even when boys and girls perform the same. Gender stereotypes and bias that impact teachers' perceptions of their students' mathematics performance, achievement, and aptitude may drive the process of differential ratings.

In general, trends suggest that as early as second grade, girls are less confident in their ability to engage in mathematics tasks and are more anxious than boys about mathematics performance (Casey, 2017; Ganley & Lubienski, 2016a; Post, 2015). These gender differences in self-perceptions are larger than actual achievement gaps, however, as highlighted by the NAEP self-concept in mathematics achievement data and other examples shared earlier in this chapter. How are mathematics teachers and teams contributing to these young students' individual beliefs? Ganley and Lubienski (2016c) suggest that girls' attitudes toward mathematics should receive greater emphasis in day-to-day classroom experiences with teachers rather than in the short, out-of-classroom experiences and interventions such as camps and after-school programs that are often developed to increase the representation of girls and women in STEM. This position leads to placing great value on the teacher-student relationship and the way this relationship can influence students' learning.

Conclusion

In our society, gender is a factor in many situations that dictates who can do what, who can learn what, who can have what, and so on. This perspective is also present in schools and classrooms. In the context of this book, it is present in the form of the mathematics gender achievement gap. When there is one instance in which girls do not receive support to achieve in mathematics in the same ways as boys do, this is one instance too many. Whether the achievement gap is real or perceived is not the issue. The fact of the matter is there are girls in our schools who are not engaging, excelling, or both to their potential in mathematics. We have an opportunity and responsibility to improve on this issue. That is the singular aim of this book.

Use figure 1.7 to determine what actions you can take to help parents and guardians do their part to close the gap.

> Recall some of the statements you suggested in figure 1.1 (page 7) regarding what teachers say about girls and mathematics. The following three actions can help you reflect on the content in this chapter as you think further about your own perspectives about girls learning mathematics.
>
> 1. Inform colleagues (or even students!) that you are reading a book to help improve your perceptions, possibilities, and priorities related to teaching mathematics to girls. Solicit others' suggestions about what you might want to think about as you read. Consider how the ideas that you hear connect with the data shared in this chapter around gender gaps.
> 2. Revisit the reflection items that you responded to at the beginning of this chapter. Has your thinking changed? Which of these items might you use or share with others, perhaps at an open school event where you can discuss your results and thinking?
> 3. Include a relevant quote from the book in your discussions with colleagues, your students, and their families.

Figure 1.7: Actions to improve parents' and guardians' perspectives about girls learning mathematics.

Reflections

Answer the following five questions independently or in your book study to further your understanding and goals related to teaching girls mathematics.

1. There may be differences in the ways boys and girls show their understanding of mathematics in the classroom, but there is no evidence of a gap that cannot be overcome with the knowledge and craft of an effective teacher. Do you agree or disagree? Why?
2. What do you hope to learn while you read this book?
3. Are there specific girls in your life who you will keep in mind as you read?
4. Did any of the statistics or research from chapter 1 surprise you?
    - If so, what resonates with you the most?
    - If not, do these research and data reinforce what you know to be true about the mathematics gender gap?
5. In your current role, what perceptions, practices, and priorities do you think are already evident and in place regarding supporting girls as learners of mathematics?

Further Reading

Breda, T., Jouini, E., & Napp, C. (2018). Societal inequalities amplify gender gaps in math. *Science, 359*(6381), 1219–1220.

Kollmayer, M., Schober, B., & Spiel, C. (2018). Gender stereotypes in education: Development, consequences, and interventions. *European Journal of Developmental Psychology, 15*(4), 361–377.

National Science Board. (2018). *Early gender gaps in mathematics and teachers' perceptions.* Accessed at www.nsf.gov/statistics/2018/nsb20181/assets/481/early-gender-gaps-in-mathematics-and-teachers-perceptions.pdf on August 22, 2018.

Pearson, N. (n.d.). Different for girls? *International Teacher Magazine.* Accessed at https://consiliumeducation.com/itm/2018/06/29/different-for-girls on August 22, 2018.

Stoet, G., & Geary, D. C. (2018). The gender-equality paradox in science, technology, engineering, and mathematics education. *Psychological Science, 29*(4), 581–593.

CHAPTER 2

Perceptions About Girls in Mathematics

All our knowledge has its origins in our perceptions.

—Leonardo da Vinci

We often hear and use the phrase, "Perception is everything." We commonly use this phrase when we are working toward clarifying our thoughts or judging a situation. What does this phrase mean to you? We typically begin perception statements with, "I think." In this chapter, we will focus primarily on perceptions of girls in mathematics and aim to answer the question, Why do perceptions about mathematics and girls learning mathematics matter—whether these are perceptions teachers, girls, or others hold? We'll first define *belief*, *bias*, and *stereotype* to provide clarity for our discussion.

Tying Belief, Bias, and Stereotype to Perception

Write your and your team's definitions in figure 2.1 for the words *perception*, *belief*, *bias*, and *stereotype*.

DO NOW

Define the following.
Perception: _____
Belief: _____
Bias: _____
Stereotype: _____

Figure 2.1: Educator definition of perception, belief, bias, and stereotype.

Consider your definitions as you read the following brief exchange between two friends.

Sarah asks, "Did you see the pretty cupcakes on the stand?"

Sandy responds, "Yes, I think they probably taste really good."

Sarah says, "Why do you think that? You haven't even tasted them."

Sandy says, "Well, they look like they taste good."

Sarah says, "I hope you are right!"

Sandy says, "I know tasty cupcakes when I see them!"

In this dialogue, Sandy has not tasted the cupcakes yet, but she uses her observations to draw a conclusion on how the cupcakes taste. Sandy is using her perceptions of the cupcakes to make that decision. This example was about cupcakes, but we even use our perceptions when the context is about people. Our perceptions influence how we treat, communicate with, relate to, and interact with others. Consider the cupcake example again. We might be influenced to only select nicely decorated cupcakes if our perception of cupcakes is based on how appetizing they look when, in fact, cupcakes that are not so nicely decorated might actually taste better. It is only when someone challenges our perceptions that a visually appealing cupcake is also tasty (by the real taste of the cupcakes) that we might realize that we should consider a different factor. Our perceptions can both benefit and mislead us, particularly when we use our perceptions to draw conclusions about people.

Now consider your own perceptions about mathematics and about girls learning mathematics. The items in figure 2.2 will help you frame your thinking and reflection as you read the chapter. If you are engaged in a book study, the items will support group discourse and debate. You may also want to record your responses to review after you read the chapter.

Respond to the following ten items.

1. What are three words that describe how you feel about mathematics?
2. Who most influenced your beliefs about mathematics?
3. In the following list, circle the word or phrase in choices a, b, and c that describes your perception about mathematics.
    a. Boring or Exciting
    b. Hard or Easy
    c. Not for you or For you
4. Have you ever said something similar to "I hate mathematics" or "I'm just not good at mathematics"? What prompted you to say this?
5. Complete this sentence: When I think about mathematics, I _____.
6. Do you think some mathematics tasks are better for boys and some are better for girls? Explain.
7. State whether you agree or disagree with the following statement and why: "Girls are not as good at mathematics as boys."
8. Of the girls in your classroom, school, district, or all these, how would you describe their attitude about mathematics?
9. Provide one statement you believe about girls' learning of mathematics.
10. What percent of girls in your class, school, district, or combination of these achieve at grade level in mathematics? Why do you think this is the case?

Figure 2.2: Educator reflections on their perceptions of girls' ability to learn mathematics.

*Visit **go.SolutionTree.com/mathematics** for a free reproducible version of this figure.*

Consider your perceptions as you read the following definitions of *belief*, *bias*, and *stereotype*.

- **Defining *belief*:** A belief is an acceptance of an idea, statement, or happening as being true. Believing something means that you commit to what is being presented. If we say, "I believe in hard work," it means that we commit to accepting that working hard is a good thing and should be done, that there is some truth that the outcome of working hard is positive. It is having confidence that working hard is worth engaging in and expecting that if no one else works hard, you will work hard because you believe in doing so. We believe certain things because of what we've learned independently, what we've been taught, what we experience, and what we've been exposed to or immersed in. We enact our beliefs through our statements and actions. In many instances, people can determine what we believe by observing what we say or do or how we behave or act. Use figure 2.3 to reflect on your beliefs about girls studying mathematics.

What is one of your beliefs about girls studying mathematics?

Figure 2.3: Educator beliefs about girls studying mathematics.

- **Defining *bias*:** To have a bias means that you have favor for something or against something, and most likely, this bias is based on a belief regarding the thing or person (or other) you favor or do not favor. Typically, to have bias means you are willing to act in such a way that what or who the bias is against is often unfair to that thing or person. Another way of thinking about a bias is to think about a preference. Often our preferences for something reflect a bias against something else. Consider the statement, "The coach only wants runners with long legs on the track team." It is evident that the coach has a bias toward runners with long legs or perhaps against a runner that does not have long legs. After coaching for a period of time, she concludes that runners with long legs always run faster than runners with shorter legs. Therefore, the coach is more likely to select runners with long legs for the team and sometimes ignoring runners with shorter legs who may be just as fast or faster. Use figure 2.4 to reflect on your bias (implicit or obvious) about girls studying mathematics.

What is one of your biases regarding girls studying mathematics?

Figure 2.4: Educator bias about girls studying mathematics.

- **Defining *stereotype*:** Let's take a look at stereotype next. A stereotype is a held thought or belief about a person or thing, but this thought or belief is often overgeneralized to others and can be very superficial. In many circumstances, people use a stereotype to classify a group of people in such a way that if one person in the group has a certain characteristic, then all the people in the group are labeled with that characteristic. So, for example, if one little boy wants to be a firefighter, a person might assume or make a statement to claim that most little boys want to be firefighters. For this simple example, it is apparent that one little boy voicing his life dreams does not speak for all little boys, but stereotyping ignores that logic. Often stereotyping assigns a role to a person because of a (sometimes misplaced) characteristic where that person is believed to be like a spokesperson for a larger group of people. For example, we might ask a person from the South to explain why people from the South drink sweet tea. First of all, the person being asked may not even drink sweet tea. Secondly, just because a person is from the South, the speaker assumes that he or she can answer for every Southerner. Use figure 2.5 to reflect on stereotypes you have regarding girls studying mathematics, and visit **go.SolutionTree.com/mathematics** for an additional activity to help you practice identifying beliefs, biases, perceptions, and stereotypes, as well as an answer key.

What is one stereotype you have about girls studying mathematics?

Figure 2.5: Educator stereotype about girls studying mathematics.

There is a relationship among beliefs, biases, perceptions, and stereotypes: our beliefs, our biases, and our stereotyping all contribute to our perceptions (Perception Institute, 2018). This is why we anchor this book on perceptions. We see perceptions at the intersection of beliefs, biases, and stereotypes. Specifically, perceptions are based on our beliefs, mediated by our biases, and reflected in our stereotypes. Based on this position, if we address perceptions, we are addressing the other terms as well. In other words, the perceptions we have about girls learning mathematics often rest on our beliefs about girls studying mathematics, mediated by biases regarding girls studying mathematics, and reflective in stereotypes about girls studying mathematics (Fennema, Peterson, Carpenter, & Lubinski, 1990; Muijs & Reynolds, 2002; Tiedemann, 2000).

Unearthing Perceptions About Girls in Mathematics

As a backdrop to discuss your perceptions about girls in mathematics, we want to examine these perceptions within three settings: (1) the classroom, (2) the school and district, and (3) the school-home connection. These are the three contexts in which our perceptions about girls studying mathematics regularly surface and for which educators have a measure of influence that can directly impact girls' experiences in mathematics. For each perspective, we propose norms that will support girls in mathematics, impacts that might lead to positive perceptions about girls in mathematics, and practical recommendations for how to incorporate the norms and impacts.

The Classroom

The classroom is the environment closest to the student, and it is where students spend the majority of the school day—in some districts, six hours per day, 180 days per year. The teachers' setting of this environment is important because what happens in the classroom is critical to students' overall school experience and can make a lasting difference in students' lives. In addition, perceptions that teachers have about girls learning mathematics can have great influence over girls' performance in mathematics. Catherine Hill, Christianne Corbett, and Andresse St. Rose (2010) specifically state, "Negative stereotypes about girls' abilities in math can indeed measurably lower girls' test performance" (p. xiv). It is our goal to promote positive perceptions about girls in mathematics and anticipate that those positive perceptions will lead to positive results. Before we do that, use figure 2.6 to share how conducive your classroom is for girls learning mathematics and detail ways your classroom promotes a positive perception of girls.

Respond to the following.
- How would someone (a parent, a guardian, a student, a colleague, or an administrator) describe your classroom in regard to its conduciveness to girls learning mathematics?
- What images, rules, procedures, and so on in your classroom promote positive perceptions of girls in mathematics?

Figure 2.6: Educator reflection on how conducive a classroom is for girls learning mathematics.

Norms

On a scale of 0 (very little) to 10 (very high), how would you rate your knowledge of mathematics? How would you rate your knowledge of mathematics for teaching? We ask these two questions because sometimes teachers' lack of confidence in their own mathematics knowledge and ability translates to students (Beilock et al., 2010). For instance, when facilitating mathematics professional development for teachers, we've often heard teachers respond to mathematics tasks or problems by saying things like, "My students could never do this" or "This would cause my students to shut down." The perception is that because it is hard for the teacher to think critically about the task or problem that it would also be hard for students to think about the task or problem. This potentially and specifically can send wrong messages to students, and of particular concern, girl students. We want to promote norms regarding the teacher's role in the classroom, particularly as it pertains to girls as learners of mathematics. The norms we propose are a collection of three norms from the *Making Sense of Mathematics for Teaching* grade-band books (Dixon, Nolan, Adams, Brooks, & Howse, 2016; Dixon, Nolan, Adams, Tobias, & Barmoha, 2016; Nolan, Dixon, Roy, & Andreasen, 2016; Nolan, Dixon, Safi, & Haciomeroglu, 2016). The teacher should establish each norm in the classroom to create a classroom environment conducive to learning.

1. **Provide explanations and justifications with solutions:** Students offer details about how and why they came to the solution they shared. In this process, students also clarify the mathematical thinking that they used in an effort to help others understand their individual or collective thinking.

2. **Make sense of classmates' solutions:** Teachers should encourage students to explain, question, and challenge one another's thinking. It is not enough to simply disagree with solutions—students should persist in understanding the explanations and justifications that their peers provide with solutions.

3. **Communicate when they don't understand or don't agree:** Rather than waiting for the teacher to facilitate a discussion of whether or not students are in agreement, students should learn how to and feel confident in expressing their questions and contrasting or conflicting ideas.

We encourage these norms because they have the potential to support positive perceptions of girls in the classroom. These norms set forth an expectation that *all* students will engage in mathematics discourse and do so in a safe and inclusive environment. In the classroom where these norms exist, girls have the space to try diverse strategies and approaches to mathematics, knowing that the teacher will acknowledge their explanations and justifications. Each norm encourages girls to try different ways of doing and thinking about mathematics, thereby fueling their confidence in themselves and their ability to do mathematics.

Impacts

How teachers interact with students influences the teacher-student relationship and how students interact with each other. How can teachers maximize that impact so that perceptions about girls learning mathematics and actions for girls learning mathematics have positive outcomes? In "Encouraging Girls in Math and Science" (Halpern et al., 2007), the U.S. Department of Education provides five recommendations for teachers regarding mathematics instruction that support achievement for all learners of mathematics, specifically considering girls. For each impact, we provide an example of how teachers can accomplish it.

1. Teach students that academic abilities are expandable and improvable in order to enhance girls' beliefs about their abilities.

    How: Sharing assessment data with students will help them see how they have improved in the process of learning mathematics and confirm to them that improvement is possible.

2. Provide students with "prescriptive, informational feedback" regarding their performance (Halpern et al., 2007, p. 9). This type of feedback should be tied to the mathematics at hand (not general).

    How: An example of specific feedback is, "I notice that you've been using the strategy 'Make a ten' in each of the examples that we have tried so far. What other strategies might you try?"

3. Expose girls to women who have excelled in mathematics or science in order to promote positive beliefs regarding women's abilities in mathematics and science.

    How: There are a variety of women in the community who teachers might invite to the classroom or who girls might shadow in the workplace for a period of time. These persons include nurses, doctors, musicians, and professors from local colleges or universities.

4. Foster girls' long-term interest in mathematics and science by choosing activities that connect mathematics and science activities to careers in ways that do not reinforce existing gender stereotypes, and choose activities that spark initial curiosity about mathematics and science content.

    How: Provide girls an opportunity to see movies like *Hidden Figures*, which tells the story of three African American women who were instrumental in applying mathematics to space flight. The movie is motivating, and it shows the outcome of using mathematics in a real-life situation.

5. Provide spatial skills training.

    How: There are many hands-on approaches for directing learners to use spatial and visualization skills. For instance, an older game but still quite captivating is the Rubik's Cube. In an attempt to arrange the cube so that each side is a single color, learners focus visually while they also use mental thinking and physical actions to solve the problem of the cube.

Which of these recommendations have you already tried? What were the results? If you have not tried any, we encourage you to select at least one to put into action during the school year.

The School and District

Perceptions about girls as learners of mathematics do not just exist in society without context. One of the strongest contexts for perceptions about girls as learners of mathematics is in schools and districts. School and district leaders and administrators (superintendents, principals, curriculum specialists, coaches, and so on) play a big role in establishing the culture of the district, of schools in the district, and of classrooms in the schools. School and district leaders have great opportunity to influence perceptions about girls as learners of mathematics.

What gets communicated within and from schools and districts is important because the perceptions of those at higher levels can permeate all other levels, even to the level of the individual student (Bottoms & Schmidt-Davis, 2010). Leaders should consider how they characterize school and district priorities and the challenges that present obstacles to girls in mathematics. Consult with fellow teachers, coaches, and school and district administrators to consider responses to these items. The answers have a great impact on how they consider and support perceptions of girls in mathematics.

Another angle to consider is how your school or district collects and uses students' achievement data. Often others use the school or district data to communicate comparisons by gender. See table 2.1 (page 26) for some outstanding research about these data and how others interpret them. We invite you to examine your own interpretations.

The Educational Research Center of America (2016), National Science Board (2016), and Halpern et al. (2007) corroborate these points. So, you probably wonder then if gender differences are not primarily about performance in mathematics, then what are they? See the four areas Ganley and Lubienski (2016a) point out in regard to perceptions: student attitudes toward mathematics, student use of strategies for problem solving, spatial awareness in relation to mathematics performance, and teachers' mindsets. (See chapter 1 on page 7 for an explanation of these four areas and research on gender differences.)

So, while schools and districts consider achievement data and how they compare across genders, it is important to keep the previous points in mind.

Norms

We want school and district leaders to be aware of and address systemic messaging that positions girls as performing lower than boys in mathematics. This messaging goes against findings that suggest that gender differences in mathematics are not completely based on aptitude. School and district leaders should think about the subtle and explicit messages they convey about girls learning mathematics. Subtle messages might include the types of images and inclusion of male and female students in mathematics texts they use, especially around contexts of word problems. Explicit messages might include how many girls they track for gifted and magnet mathematics. To support girls in mathematics, we propose two norms at the school and district levels.

Table 2.1: Interpretations of Research on Gender Differences in Mathematics Performance

Research Finding	Possible School or District Interpretation
Ganley and Lubienski (2016a) report small mathematics differences in performance. Writer Beth Azar (2010) cites experts (Ceci & Williams, 2010; Else-Quest, Hyde, & Linn, 2010) who agree that any presence of greater achievement among one gender is small. Hill and colleagues (2010) say that differences in performance disappear when test administrators tell students that girls and boys are equally capable in mathematics.	Not all research supports a conclusion that boys consistently outperform girls in mathematics.
Ganley and Lubienski (2016a) state, "Gender differences on math tests tend to be more pronounced when the content of the assessment is less related to the material that is taught in school (for example, on the SAT-Math as opposed to a math test in school)." Professors of economics Muriel Niederle and Lise Vesterlund (2010) state that "competitive pressure may cause gender differences in test scores that exaggerate the underlying gender differences in math skills" (p. 140).	We may see more of what we believe to be gender differences in mathematics on assessments that are not classroom based.
Ganley and Lubienski (2016a) point to research that finds "that gender gaps are larger among higher-performing students." Glenn Ellison and Ashley Swanson (2010) state that "the gender gap on math tests among high-achieving students is consistently much larger" (p. 109), with examples of data from the mathematics SAT and Program for International Student Assessment (PISA) as support.	Gender differences in mathematics are not often revealed among lower-performing students.

*Visit **go.SolutionTree.com/mathematics** for a free reproducible version of this table.*

1. **Have an authentic goal for communicating data that are gender specific:** This is important because we know that differences in mathematics performance by gender do not necessarily signify that girls are not capable of excelling in mathematics.

2. **Provide support for all students, but particularly girls to address factors beyond aptitude to help girls succeed in mathematics:** Specifically, promote instruction that does four things.
   a. Helps girls develop positive attitudes toward mathematics
   b. Encourages girls to use inventive strategies for problem solving
   c. Supports girls' growth in spatial awareness
   d. Comes from female teachers who are confident in mathematics

Refer to figure 2.7 to consider your school or district achievement.

 Access achievement data from your school or district. Consider any data that employ gender as a factor. What is the messaging of these data?

Figure 2.7: Analyzation of school or district data for messaging.

Impacts

School and district leaders can make a difference in how perceptions of girls in mathematics develop. The following are three ways that impact can occur.

1. Respond to the outcome of any schoolwide or districtwide mathematics assessment data that indicate girls are underperforming by examining potential elements that could undermine girls' ability to perform well in mathematics.

    How: One simple way to find out what might be hindering girls' success in mathematics is to ask the girls. Here are three sample questions that will solicit insight from girls about their mathematics experiences.

    a. What do you need to better engage or participate in mathematics class?

    b. Do you believe mathematics homework is helpful for you? Why or why not?

    c. If you do well in mathematics, why do you think so? If you don't do well in mathematics, why do you think so?

2. Establish principles that support girls studying mathematics. Because research shows that girls will often adapt to the anxiety level of female teachers, provide the professional development support that women (and all) teachers need to have confidence in mathematics.

    How: Seek out a high-quality professional development provider to collaborate with your school or district to deliver collaborative, sustainable mathematics professional development.

3. Be proactive in checking girls' attitude toward and confidence in mathematics.

    How: Take note of what students say and how they act regarding mathematics, and then move to make changes to improve any negative attitudes. For example, if girls say, "I hate mathematics. I'm just not good at it," perhaps respond by asking, "Well, what's something you love?" or "What's something you are good at?" Then make a connection between the girl's responses and mathematics. In most instances, there is a way that mathematics plays a role in things students enjoy doing.

The School-Home Connection

The home is the primary place where students develop in the preschool years, both as individuals and through their early experiences in developing number sense. Learning opportunities through activities such as playing (blocks, drawing) and household chores (dishes, laundry, shopping) impact students' mathematical development, whether formal or informal, intentional or unintentional. There is abundant evidence of the role of the home in students' academic achievement (Duncan et al., 2007; Hill et al., 2010;

Overdeck, 2018). Schools cannot realize the goal of empowering positive perceptions about girls as learners of mathematics without including contributions from students' home life. Involving *parents* (a term we use to also include grandparents, foster parents, and other guardians) in the quest to present positive influences on girls in mathematics is critical support for girls. Consider the perceptions of two parents of girls: the first parent relays the importance learning mathematics at high levels plays in her child's life and the second illustrates how the school-home connection helped her daughter's mathematics achievement.

Parent Diana Elysee reflects on the importance of learning mathematics for her daughter Jelina:

> Mathematics is one of those subject areas that have multiple ways to answer problems and discover a working response. There is not just one way of solving problems. Jelina's success in math correlates to having multiple perspectives of solving a problem. It is my desire that my daughter is successful in math to obtain the knowledge of becoming a problem solver and working on any problem in different ways with a unique perspective than just one ordinary way. Being able to solve problems will also provide her with the confidence she needs to be successful. In addition, as a problem solver, she will also flourish and thrive in a career she chooses to pursue. (D. Elysee, personal communication, October 7, 2017)

Parent Lesa Smithies, whose daughter Emily is in twelfth grade, reflects that she always loved reading and encouraged her daughter in that subject. While Emily was an avid reader, she also exhibited a gift for interacting with numbers and mathematical concepts. Smithies said:

> As I reflect on it now, I believe my own hostile feelings about math were interfering with what I was hearing. I continued to foster her love for reading because that was my area of comfort. . . .
>
> I believe Emily was inclined to do well in math, but she has had great teachers who developed that love and continued to inspire her to keep it going. I am extremely appreciative of this as I didn't do it for her because of my own reservations about math. I will be forever grateful that others were able to see a skill that at first I didn't. I believe we are still in a male-driven society, so it's very important to have someone advocating and pushing young women to succeed. My hope is that my daughter would be one of those that not only inspires but opens the door for other young women in the field of mathematics. (L. Smithies, personal communication, October 9, 2017)

You see the role parents' perceptions play in their daughters' success. With Emily's mother, for example, you see her belief that a key factor in the path to Emily's success was the support of powerful teachers of mathematics throughout Emily's education. Beyond the role of providing instruction in the mathematics classroom, teachers, such as those who worked with Jelina and Emily, can serve as role models and advocates for the advancement of mathematical opportunities for girls.

Use figure 2.8 to list what you've heard girls' parents say about mathematics and the daughters' ability to learn the subject and to explain what you're doing to support the school-home connection.

Norms

When parents and guardians are in partnership with teachers, great things can happen because students receive support for learning at home as well as at school. To support the school-home connection, educators can adopt two norms.

1. **Establish relationships with parents as a partnership to support girls as learners of mathematics:** Teach parents about the relevant mathematics content for your grade level.

Respond to the following.

- What have you heard parents or guardians say about their perceptions of girls as learners of mathematics? What do mothers, grandmothers, or other women in your students' lives say about their own abilities? Ask them!
- What are three things you are already doing to facilitate school-home connections for your students, particularly around their learning of mathematics?

Figure 2.8: Reflections on girls' perception of mathematics and their ability to learn mathematics.

Give them practical suggestions about ways to support their children in mathematical experiences, including those related to school (homework support) and those that are more informal (discussions about family budgeting, shopping, and organizing).

2. **Recognize possibilities where parents can be resources for improving the perceptions of girls in mathematics and in helping girls improve perceptions of themselves as learners of mathematics:** Teach parents about possibilities to help their children overcome negative or neutral self-concepts in mathematics, including language and discussions about the nature of learning (see the direct messages adults can use in Student Confidence Levels Impacting Mathematics Achievement, page 15).

In employing the first norm, teachers create more opportunities for girls to achieve and receive motivation to study mathematics:

> When teachers and parents tell girls that their intelligence can expand with experience and learning, girls do better on math tests and are more likely to say they want to continue to study math in the future. That is, believing in the potential for intellectual growth, in and of itself, improves outcomes. (Hill et al., 2010, p. xiv)

With the second norm, family members might have unique occupations and other ways of doing mathematics that can provide a personal connection to mathematics for girls (as well as boys). Furthermore, mothers and other women relatives of girls might serve as great role models for women using mathematics in everyday life as well as in career opportunities. It is important to build relationships that support school-home connections that benefit classroom instruction. Use figure 2.9 to brainstorm steps you can take to strengthen this connection.

What action can you take in the near future to strengthen the school-home connection for the girls in your classroom so that they receive additional support to excel in mathematics?

Figure 2.9: Ideas to strengthen school-home connection.

Impacts

School-home connections have great potential for influencing girls to study mathematics (Wiest, 2014). One important thing to do is honoring students' families, culture, background experiences, and so on as resources for students' learning. In some instances, it requires engaging families in the mathematics learning experience so that they better understand what and how students are learning mathematics in the 21st century.

To use the school-home connection to build positive perceptions about girls in mathematics, consider the following three methods.

1. Be explicit with families.

   How: Tell them how important it is for girls to experience their families having positive perceptions about mathematics and positive perceptions about girls studying mathematics.

2. Use mathematics confidence-building exercises that students can share with families at events such as a school open house or parent-teacher meetings.

   How: For example, play games (such as Mancala and board games that include numbers) that have mathematics as a context to help girls and their families experience fun and joy while using mathematics as a strategy to win at the games.

3. Include ideas about mathematics that students bring from home.

   How: Do this rather than trying to force students to dismiss the way their parents show them how to do mathematics. This kind of response to parents only sets up a wall between school and home.

Conclusion

Examining your own perceptions of girls as learners of mathematics is one early step in planning for shifts in your classroom instruction and seeking ways to empower your girl students. You've already taken a big step by starting to read this book. As you continue to read, it may be helpful to revisit your notes about your perceptions and to reflect on how they may evolve as you learn more about how to make sense of mathematics for teaching girls.

Applying efforts to make sure your classroom, school, district, or all of these are alert to research findings on the mathematics achievement of girls will be helpful when having conversations about norms that support girls' learning of mathematics. Of course, it is always important to consider your own local context as you review generalized research. Overall, the goal is to provide every opportunity for girls to be successful in mathematics and to use various models to do so. See figure 2.10 for a list of actions to build the self-confidence of girls in your class.

> The following three actions can help you build self-confidence for girls in mathematics.
>
> 1. Consider programs, initiatives, incentives, or combinations of these around mathematics learning that take place at your school. How do they represent girls in these examples? How might girls perceive these opportunities based on how they are introduced or how they have occurred? How might you increase the participation and representation of girls?
>
> 2. Develop a confidence-building activity (like Mancala) you might use with your girl students. For example, make sure to balance opportunities for students to demonstrate their knowledge so that girls learn just as well as boys to construct arguments, justify and explain their thinking, and take risks for learning in a safe environment. In reading, some teachers use an author's chair, where they select a student to sit and share his or her stories with the class and to receive feedback. The author then has support and confidence for improving his or her story. Why not establish a mathematician's chair? Students can sit in the chair and share their mathematics solutions and receive feedback and gain strength and confidence in the same fashion.
>
> 3. Talk to your girl students about how they feel about themselves as learners of mathematics. For example, during a mathematics activity (formal or informal), ask one how she is feeling about her progress. Offer support to build confidence by praising her effort to persevere, to try a different strategy, or to represent her thinking in unique ways.

Figure 2.10: Actions to build self-confidence for girls in mathematics.

Reflections

Answer the following five questions independently or in your book study to further your understanding and goals related to teaching girls mathematics.

1. What structures are currently in place in your district or school to positively impact girls' perceptions of themselves as learners of mathematics?

2. How do you engage students so that you are helping them to develop positive perceptions about girls studying mathematics?

3. What is a summary of a conversation you've had with a parent about his or her daughter studying mathematics?

4. Consider this quote from Christianne Corbett and Catherine Hill (2015): "We all hold gender biases, shaped by cultural stereotypes in the wider culture, that affect how we evaluate and treat one another." What can you say to confirm or refute it?

5. What is something you will say or do during your next conversation with a female student to convey your belief in her as a learner of mathematics?

Further Reading

Hyde, J. S., Canning, E. A., Rozek, C. S., Clarke, E., Hulleman, C. S., & Harackiewicz, J. M. (2017). The role of mothers' communication in promoting motivation for math and science course-taking in high school. *Journal of Research on Adolescence, 27*(1), 49–64.

Leonard, J. (2008). *Culturally specific pedagogy in the mathematics classroom: Strategies for teachers and students.* New York: Routledge.

Levine, G. (2013, October 25). *Closing the gender gap: Increasing confidence for teaching mathematics.* Proceedings from the 44th Annual Conference of the Northeastern Educational Research Association, Rocky Hill, Connecticut. Accessed at https://opencommons.uconn.edu/cgi/viewcontent.cgi?article=1006&context=nera_2013 on August 28, 2018.

Wang, M. T., & Degol, J. L. (2017). Gender gap in science, technology, engineering, and mathematics (STEM): Current knowledge, implications for practice, policy, and future directions. *Educational Psychology Review, 29*(1), 119–140.

CHAPTER 3

Possibilities for Girls in Mathematics

Education is a natural process carried out by the child and is not acquired by listening to words but by experiences in the environment.

—Maria Montessori, Physician and Educator

Many educators have taken a number of actions to encourage girls in STEAM (science, technology, engineering, arts, and mathematics) studies and careers. For example, as a former elementary educator and assistant principal at Idyllwilde Elementary in Sanford, Florida, Keith Erickson knows how important it is to illustrate the infinite possibilities for girls in mathematics. He and his staff (K. Erickson, personal communication, December 3, 2017):

- Worked with grades K–5 students to engineer models of bridges, roller coasters, and rockets
- Founded an in-school academic club program, STEAM Teams, in which every teacher focuses on either science, technology, engineering, arts, or mathematics for thirty minutes

Erickson was inspired to take these actions after hearing Girls Who Code founder Reshma Saujani share in a 2016 keynote that within five years of her speech 1.4 million computer-related jobs would open in the United States and women would fill only 3 percent of them:

> I thought, "What? How is [that] possible? What can we do to fix this?" . . . Not every girl or boy is going to be interested in mathematics or engineering, but if we don't expose them, encourage them, and teach them STEM topics, how will they ever know? Girls can code. Girls can engineer. Girls can do math. They can do these things just as good as boys and it is time that educators show them that they are just as good. (K. Erickson, personal communication, December 3, 2017)

In figure 3.1, share your own reaction to this statistic.

Respond to the following.
- What were your reactions to the statistic that women will occupy 3 percent of computer-related jobs in 2021?
- Have you implemented any strategies or programs similar to those Keith Erickson used for your classroom, school, or district?

Figure 3.1: Educator reaction to the number of women professionals in computer-related roles.

In this chapter, we will focus on eight possibilities for girls in mathematics and discuss planning for positive practices in regard to these possibilities. By *possibilities*, we mean the actions that teachers might engage in to prepare and to facilitate opportunities for girls to learn mathematics effectively. We'll answer the question, What *possibilities* of actions and activities might promote girls' success in mathematics? As, you consider this question, ask yourself how you meet the needs of girls in your classroom so that they leave as confident students who make sense of mathematics. Teaching practices, particularly in the early years, impact the future mathematics success of the students. As outlined in chapter 1 (page 7), there is mixed research on the presence of a gender gap in mathematics (Beilock et al., 2010; Fennema et al., 1998; Ganley & Lubienski, 2016a; Tzuriel & Egozi, 2010). But one thing is undeniably true: the educator plays a critical role in student success. Taking time for reflection plays directly into that success.

As an educator, it is important for you to be reflective. Look closely at what you do in the classroom and how it impacts the students you are teaching. So much time is spent planning and implementing lessons, often educators forget about taking the time to reflect on the successes and struggles of a lesson and individual learners. Take a few moments to reflect on the items in figure 3.2 as they relate to your teaching of mathematics.

> Respond to the following seven items as they relate to your attitude about teaching, girls as learners of mathematics, and your practices.
>
> 1. Do you enjoy teaching mathematics? Why or why not? What do you think your students would say about your response to this question?
>
> 2. Have you ever noticed a difference in the way the girls in your class learn mathematics compared to the boys? If so, what did you notice?
>
> 3. State whether you agree or disagree with the following statement, and explain your reasoning: "The only way to differentiate in my classroom is through ability level."
>
> 4. Complete this sentence: After an assessment, either formative or summative, I provide feedback to my students by _____.
>
> 5. Describe how you organize your mathematics block. What percentage of the block do you spend with direct instruction from the teacher, students talking, independent work, and group work?
>
> 6. When your students are problem solving, what tools are available for them to use?
>
> 7. What are some extracurricular activities or clubs your school has available for students? Are boys and girls equally represented in mathematics-based activities and clubs?

Figure 3.2: Educators reflect on mathematics teaching practices, tools, and programs.

Visit **go.SolutionTree.com/mathematics** *for a free reproducible version of this figure.*

Think about your answers to the questions in figure 3.2. You may think you already offer many opportunities for girls learning mathematics, but the best teachers are those who continually try to improve their teaching by reaching the needs of all students. As we establish positive teaching practices for girls studying mathematics, we want to look at teaching practices from different possibilities. Positive teaching practices can impact all students' future success, with a specific focus on empowering girls. In order to improve our practices in the classroom so that girls excel in mathematics, we present eight possibilities to consider.

1. Honoring diverse ways of doing mathematics
2. Fostering classroom discourse

3. Planning for hands-on learning
4. Using questioning to boost understanding
5. Using formative assessment
6. Considering contexts for tasks
7. Modeling of mathematical power
8. Conveying teacher expectations

We will take an in-depth look at each possibility by exploring what the research states and offer a shared vision for teaching mathematics through classroom video clips that share an example. You'll notice that we will use the following reproducible guiding questions (figure 3.3) to frame both your lens for viewing the classroom videos and the debrief that is offered for each example.

> Answer the following three guiding questions while watching classroom video examples.
> 1. What actions does the teacher take to engage the girls in learning mathematics?
> 2. What is the evidence that the girls are actively engaged in learning mathematics?
> 3. How does the teacher challenge the girls to think critically about the mathematics?

Figure 3.3: Guiding questions.

*Visit **go.SolutionTree.com/mathematics** for a free reproducible version of this figure.*

We will also make connections between each possibility and the applicable Standards for Mathematical Practice from the National Governors Association and the Council of Chief State School Officers (NGA & CCSSO, 2010). These eight Mathematical Practices are applicable across all grade bands and are related for any mathematics curriculum.

1. Make sense of problems and persevere in solving them.
2. Reason abstractly and quantitatively.
3. Construct viable arguments and critique the reasoning of others.
4. Model with mathematics.
5. Use appropriate tools strategically.
6. Attend to precision.
7. Look for and make use of structure.
8. Look for and express regularity in repeated reasoning.

After explaining the eight possibilities, we explore their role in planning for positive teaching practices within three contexts: (1) the classroom, (2) the district, and (3) the school-home connection. In the course of doing so, we will provide a teaching tool that is ready for application in the classroom for each of the eight possibilities.

Honoring Diverse Ways of Doing Mathematics

It is not an unusual notion that students learn and do things in different ways. In fact, we would expect that diverse ways of doing mathematics will be present because there is commonly more than one way to work on a task or solve a problem. Therefore, we should not take the position that girls have to do mathematics tasks in the same way that boys do mathematics tasks. When any learner, regardless of gender, does mathematics tasks in a different way, we should honor his or her diverse way of doing and thinking about mathematics; however, we want to foster the conceptual understanding and flexible thinking of all learners based on widely adopted content and process standards in mathematics. In *Good Questions: Great Ways to Differentiate Mathematics Instruction*, Marian Small (2009) suggests that "one way that we see the differences in students is through their responses to the mathematical questions and problems that are put to them" (p. 2). She further proposes that it is important "to ensure that each student in the class has the opportunity to make a meaningful contribution to the class community of learners" (Small, 2009, p. 3).

A valuable element of honoring diverse ways of doing mathematics is recognizing that when girls "make sense of problems and persevere in solving them" (Mathematical Practice 1; NGA & CCSSO, 2010), girls will often have perspectives that are different from their peers and may even be different than the teacher's. Consider how the teacher honors diverse ways of doing mathematics that take place in this fifth-grade class video, Finding the Volume of Rectangular Prisms, and respond to the guiding questions in figure 3.3 (page 35). (Visit **go.SolutionTree.com/mathematics** to download a reproducible version of the questions.)

Finding the Volume of Rectangular Prisms:
www.SolutionTree.com/Finding
_Volume_of_Rectangular_Prisms

Now that you had the opportunity to observe this fifth-grade lesson on finding the volume of rectangular prisms, consider how the teacher honors diverse ways of doing mathematics through her task selection and facilitation of the lesson. We now provide additional insight via our reactions to the guiding questions.

What Actions Does the Teacher Take to Engage the Girls in Learning Mathematics?

In the video, the students each receive the opportunity to find multiple ways to build a box using a total of thirty-six cubes. By recording a variety of student approaches and encouraging students to talk with one another about their observations, the teacher validates each student's contribution while also building on the class's collective knowledge of how to determine the volume of rectangular prisms in this exploratory lesson. During her questioning at individual tables where students work, the teacher encourages students to further explain what they notice as they work and to link their observations to the procedures that help them identify the volume of rectangular prisms.

What Is the Evidence That the Girls Are Actively Engaged in Learning Mathematics?

As the video unfolds, the students in this classroom are actively engaged in learning mathematics through several sources of evidence. We can see that students are using the manipulatives provided for this open-ended task in order to find several possible solutions. As individual students offer their solutions for the teacher to record on the board, it becomes apparent that students have to further recognize whether the additional solutions that they identify are similar to or different from those that their classmates have already shared. Students also receive time to analyze their solutions and the solutions of others in order to determine the relationship between the length, width, and height of the rectangular prisms that they build in the first part of the task.

How Does the Teacher Challenge the Girls to Think Critically About the Mathematics?

After students share their initial responses to the box-building task, their teacher asks them to talk with each other about the examples that they have recorded on the board, in addition to the other examples that they may have identified. The teacher uses open-ended questions such as, "What do these numbers mean?" and "What do you notice?" to challenge students to extend their thinking beyond the initial task in order to identify the formula for calculating the volume of a rectangular prism. This instructional approach, anchored in the selection of a rich task paired with effective, facilitative questioning, supports the students to do the sense making, rather than supplying the students with the formula and offering opportunities for practice and application.

Use figure 3.4 to think about multiple strategies your students use to solve mathematics problems.

Think about a time in your classroom when students used multiple strategies to solve the same mathematics problem. What were some of the strategies that you used to shape the discussion and your students' understanding of the content?

Figure 3.4: Reflection on strategies students use to solve a mathematics problem.

Fostering Classroom Discourse

Classroom discourse (most typically thought of as oral communication) provides an opportunity for girls to share their thinking and talk about mathematics. This talk might involve sharing explanations, justifications, strategies, and so on. According to Noreen M. Webb, distinguished professor of Social Research Methodology at University of California, Los Angeles, and her colleagues (2017), both the giving and the receiving of information in the midst of classroom discourse benefits learners. The importance of classroom discourse is not new in mathematics. Massey University professors Margaret Walshaw and Glenda Anthony (2008) conducted a review of research on the topic of classroom discourse. One of their most pressing findings is that "the effective use of classroom discourse makes students' mathematical reasoning visible and open for reflection" (p. 539). A valuable element of classroom discourse is girls learning how to do Mathematical Practice 3, "Construct viable arguments and critique the reasoning of others" (NGA & CCSSO, 2010). Girls might find discomfort in what appears to be a competitive stance. However, the focus is not on girls competing with boys or even other girls. It is on providing a safe space for girls to share their mathematical thinking and to express their reasoning as they build understanding.

One way to support classroom discourse is to select tasks that you anticipate will elicit varied (and perhaps conflicting) responses and plan for facilitation that will engage learners to explain their own reasoning and the reasoning of others (Mathematical Practice 3; NGA & CCSSO, 2010). Consider how the teacher promotes classroom discourse in the video clip of a third-grade class defining and classifying squares and rectangles, and respond to the guiding questions in figure 3.3 (page 35).

Defining and Classifying Squares and Rectangles:
www.SolutionTree.com/dixon-k-2/defining-and
-classifying-squares-and-rectangles

Now that you had the opportunity to observe this third-grade lesson focusing on defining and classifying squares and rectangles, consider how the teacher promotes classroom discourse through her task selection and lesson facilitation. We now provide additional insight via our reactions to the guiding questions.

What Actions Does the Teacher Take to Engage the Girls in Learning Mathematics?

The teacher asks students to make squares and rectangles, offering explanations about how to name and classify shapes. The teacher uses the following actions to engage the learners in mathematics.

- **Offering opportunities for students to use manipulatives to represent their thinking:** This provides access for students to pair abstract thinking with concrete or representational models.

- **Using follow-up questions and opportunities for students to change their solutions and share their thinking as they identify their own errors and misconceptions:** This supports all students to stay engaged in collective discourse as they are making sense of the task presented and one another's thinking, thereby solidifying their own understanding.

- **Offering an alternative approach to solving the task that challenges students' thinking and promotes discourse (both in small groups and the whole class):** This introduces an opportunity for students to deepen their understanding, while making sense of the mathematical thinking of others (whether the alternative approach is real or imagined).

- **Asking students to share whether they agree or disagree with one another:** This helps to build the social norm that students must listen carefully and explain and justify their mathematical thinking with one another.

- **Directly addressing (and in this case, correcting) students' misconceptions with the intent to promote additional discussion about the content:** This offers the teacher an opportunity to uncover potential hidden misconceptions or gaps in conceptual understanding.

What Is the Evidence That the Girls Are Actively Engaged in Learning Mathematics?

The girls in this lesson are actively engaged, with evidence based on the students' constructed models of shapes using the manipulatives provided. Students' explanations and justifications, in response to the presented task and the teacher's follow-up questions, also offer additional evidence of students' active engagement with the geometry content in the lesson. As many students share explanations that are not accurate (based on their misconceptions around the defining attributes of a rectangle), the teacher uses the opportunity to directly address the misconception paired with discussion at student tables to support students' understanding of rectangles, squares, and the classification of each.

How Does the Teacher Challenge the Girls to Think Critically About the Mathematics?

There are opportunities for the teacher to directly challenge the thinking of girls during the lesson, causing them to think critically about the geometry content at hand. For example, the teacher gives students the task to construct a square in the first task. Then, the teacher identifies the example that one of the girls in the class offers that is not a square (because it does not have equal sides). The use of follow-up questioning, time for the student to self-identify her error, and the opportunity for the student to explain what she could do to change her model in order to accurately represent a square all exemplify how the teacher challenges this particular student to think critically. We see an additional example when the teacher asks another girl to make a square that is not a rectangle, a follow-up to a misconception that the student offers. This responsive teacher move allows the girl student (and likely others in the room) to realize that it is not possible to make a square that is not a rectangle, deepening her understanding of the attributes and classification of shapes.

Use figure 3.5 to list ways to facilitate positive, productive discourse in your classroom.

How do you facilitate positive and productive discourse in your classroom?

Figure 3.5: Ways to facilitate positive and productive discourse in mathematics class.

Planning for Hands-On Learning

Hands-on learning (such as, use of manipulatives and other instructional tools) provides an opportunity for girls (and all learners) to actively engage during instruction. This active engagement creates a bridge between students' concrete understanding and abstract understanding. Boise State University associate professor of educational technology Lida J. Uribe-Flórez and Virginia Tech University professor Jesse L. M. Wilkins (2017) state that "manipulatives provide students with opportunities to visualize and maneuver abstract mathematical concepts in concrete ways" (p. 1542). Since we know that visualization is one strength that students need to develop to have success in mathematics, our aim is to employ hands-on learning to improve students' abilities to visualize mathematical concepts. In addition, "a good manipulative bridges the gap between informal math and formal math and formal school math" (Smith, 2009, p. 20). Uribe-Flórez and Wilkins (2017) further conclude that the more elementary students use manipulatives, the more they learn in mathematics—a key to building girls' confidence.

A valuable element of hands-on learning is girls accomplishing Mathematical Practice 5, "Use appropriate tools strategically" (NGA & CCSSO, 2010). (Note that Mathematical Practice 5 assumes that teachers make appropriate tools available to students and subsequently those students learn to use the tools in an effective manner.) Consider the hands-on learning that takes place in this video of a second-grade class solving a problem on place value, and respond to questions in figure 3.3 (page 35).

Introducing Place Value With the Candy Shop:
www.SolutionTree.com/Introducing_Place
_Value_With_the_Candy_Shop

Now that you had the opportunity to observe this second-grade lesson on place value, consider how the teacher supports the use of manipulatives through her task selection and facilitation of the lesson. We now provide additional insight via our reactions to the guiding questions.

What Actions Does the Teacher Take to Engage the Girls in Learning Mathematics?

In this classroom video, the teacher presents students with a task that has a single, correct answer, yet the possibility for multiple strategies. She gives manipulatives to each table (rather than to individual students or partners) with the intent to promote student interaction, discourse, and opportunities to critique the reasoning of one another in small working groups. As the lesson unfolds, the teacher identifies opportunities to use students' various responses as a springboard for meaningful discussion about the task, while also addressing misconceptions about place value that emerge as they are counting the cubes in the boxes at their tables.

What Is the Evidence That the Girls Are Actively Engaged in Learning Mathematics?

In the video, we see two girls who initially disagree about how many pieces are in the box: one student knows that there are one hundred pieces because there are ten rows of ten pieces, but the other student thinks that there might be sixty or seventy based on her estimate from looking at the snap cubes as they are arranged in the box. Through the use of hands-on learning and a strategic tool to represent place value, the two girls and their group members come to a consensus that is accurate. Perhaps even more important, the girls explain and justify their answer in order to help themselves and their group members make sense of the problem. This brief example highlights the impact of manipulatives and how teachers can use them in contexts of hands-on learning to strengthen girls' ability to represent and solve problems and develop the strategy of visualizing. This in turn will support advanced mathematical concepts and procedures.

How Does the Teacher Challenge the Girls to Think Critically About the Mathematics?

As the teacher engages in dialogue with students at their table groups, we hear her ask questions such as, "How can you help her to make sense of what you just said?" and "What does she mean by that?" These facilitative questions help students make sense of their own thinking and the thinking of their classmates. They also enable the teacher to extend students' thinking when she intentionally connects students whose strategies or solutions are not the same as one another. For example, during the whole-group discussion, we hear multiple solutions ("100," "124," and "106") when the teacher asks students to share whether they agree or disagree with one another and to explain and justify their thinking. She then asks content-specific follow-up questions (like "Why would she count by tens?") that encourage students to use manipulatives in their response. Doing this allows students to recognize their errors, revise their answers, and explain their new problem-solving process.

Use figure 3.6 to share how you use manipulatives in your classroom and to relate the strengths and barriers to incorporating academic tools in the classroom.

Respond to the following.
- Do you utilize manipulatives in your classroom currently?
- What are the strengths and barriers to incorporating academic tools in the classroom? How do the strengths outweigh the limitations off manipulatives?

Figure 3.6: Review of how teachers use manipulatives in the classroom.

Using Questioning to Boost Understanding

In *Making Sense of Mathematics for Teaching Grades K–2* and *Making Sense of Mathematics for Teaching Grades 3–5* (and the secondary grade-band books as well), the authors take the position that "teachers who have a deep understanding of the content they teach facilitate targeted and productive questioning strategies because they have a clear sense of how the content progresses within and across grades" (Dixon, Nolan, Adams, Brooks, & Howse, 2016; Dixon, Nolan, Adams, Tobias, & Barmoha, 2016, p. 6). Questioning facilitates classroom discourse, uncovers students' understanding, engages students in active thinking, and supports students' engagement in the Mathematical Practices. Thus, the desire to engage students in critical thinking about and with mathematics, rather than a sole focus on leading the student to producing a correct response, drives targeted and productive questions. Joan Buchanan Hill (2016), head of school for the Lamplighter School in Dallas, Texas, provides further clarification on the importance of questioning by stating "questions are central to learning" (p. 660). Hill (2016) goes on to say:

> Questioning is one of the basic techniques that a teacher can use to stimulate thinking, learning, and class participation. Its effective use engages students in authentic and meaningful ways by motivating and keeping students on tasks. Questions also focus the students' attention on a lesson's key objectives. (p. 661)

Using questions to boost student understanding in the classroom offers students opportunities to share how they are making sense of problems (part of Mathematical Practice 1) and reasoning abstractly, quantitatively, or both (Mathematical Practice 2) about the task at hand (NGA & CCSSO, 2010). Effective questioning also elicits students' explanations and justifications about their use of mathematics, which supports their ability to construct viable arguments (part of Mathematical Practice 3; NGA & CCSSO, 2010). Consider the use of questioning that takes place in this video clip of a grade 1 class working on a word problem, and respond to the guiding questions in figure 3.3 (page 35).

 Solving a Word Problem Where the Change Is Unknown:
www.SolutionTree.com/Solving_a_Word_Problem
_Where_the_Change_Is_Unknown

Now that you had the opportunity to observe a first-grade lesson focusing on a word problem where the change is unknown, consider how the teacher promotes classroom discourse through her task selection and facilitation of the lesson. We now provide additional insight via our reactions to the guiding questions.

What Actions Does the Teacher Take to Engage the Girls in Learning Mathematics?

In this lesson, the teacher intentionally selects a task in which the change is unknown in anticipation that many students may have difficulty keeping the context of the task in mind as they use their selected strategy or strategies to identify a solution. Some of the questions the teacher uses to help keep the students' focus on the context of the problem, while engaging in the mathematics to determine a solution, include:

- "I see . . . and so our answer is what?"
- "So, the answer to the problem is fifteen?"
- "So, our question was, How many more stickers does he need to have fifteen stickers altogether? What do you think?"
- "How do you know?"
- "What's eight?"
- "And so how many more stickers did Stefan need?"

To enhance instruction that uses questions to boost girls' understanding in mathematics, consider using assessing and advancing questions. In the lesson, the teacher uses a combination of advancing and assessing questions. For instance, "I see . . . and so our answer is what?" is an assessing question because it solicits what the student is thinking or what the student understands about the matter at hand. "How do you know?" is an advancing question. It is not simply asking for an answer to the task, but it is asking for a justification that will support the answer. To respond, the student has to consider whether the answer meets the criteria of the task. This is productive problem-solving behavior that will help the student beyond the task in focus.

What Is the Evidence That the Girls Are Actively Engaged in Learning Mathematics?

In this lesson, students use a variety of strategies to solve the change unknown problem (7 + ___ = 15), including counting by ones, make a ten, doubles plus one, and so on. You can observe the various strategies students use as they work at their table groups and through the shared examples that students offer during whole-group discussion. The teachers' use of questioning in this lesson focuses both on the mathematics and strategies applied and the frequent revisiting of the context to ensure that students understand how their solutions represent the change that is unknown in the word problem.

How Does the Teacher Challenge the Girls to Think Critically About the Mathematics?

In this lesson, we see the teacher challenge the girls to think critically as she asks them to explain and justify how they know that their strategy works and that they understand how their solution makes sense in relation to the task presented. The teacher asks students to not only share their own strategies, but to explain how and why the strategies their classmates use are accurate (such as the case when a girl student has to explain how her classmate's make-a-ten strategy works).

Teachers can use figure 3.7 to brainstorm how they might use assessing and advancing questions in a future lesson.

 Consider an upcoming lesson. How might you use assessing and advancing questions to engage your students?

Figure 3.7: Ideas for using assessing and advancing questions in a future lesson.

Using Formative Assessment

According to W. James Popham (2008), "Formative assessment is a planned process in which assessment-elicited evidence of students' status is used by teachers to adjust their ongoing instructional procedures or by students to adjust their current learning tactics" (p. 6). Hence, one can see that formative assessment is much more than grading and testing, but it is a process that undergirds planning with intentionality for effective instruction. NCTM (2013) describes the role of formative assessment in the classroom:

> Through formative assessment, students develop a clear understanding of learning targets and receive feedback that helps them to improve. In addition, by applying formative strategies such as asking strategic questions, providing students with immediate feedback, and engaging students in self-reflections, teachers receive evidence of students' reasoning and misconceptions to use in adjusting instruction. (p. 1)

Two ways that formative assessment differs from traditional summative assessment (for instance, yearly standardized assessments) are that it requires the teacher to commit to adjusting instruction as a result of student engagement and that it encourages students to pay attention to what they are learning and adjust their behavior to improve learning.

The process of formative assessment can have immediate impact on girls' engagement in mathematics as teachers purposely include all students in the classroom and make certain that girls receive the opportunity to perform Mathematical Practice 3, "Construct viable arguments and critique the reasoning of others" (NGA & CCSSO, 2010). Consider how the teacher uses formative assessment to gather information about girls' mathematical thought processes to boost girls' understanding in the video clip of a second-grade small group performing multidigit addition, and respond to the guiding questions in figure 3.3 (page 35).

 Add Three-Digit Numbers With Regrouping:
SolutionTree.com/GR2ThreeDigit

Now that you had the opportunity to observe a snapshot from this second-grade, small-group lesson on multidigit addition, consider how the teacher uses strategies of formative assessment to guide her lesson facilitation. We now provide additional insight via our reactions to the guiding questions.

What Actions Does the Teacher Take to Engage the Girls in Learning Mathematics?

In the video, we see the teacher presents the task and the option of tools for students to use as the lesson begins. You may notice that the teacher does not provide any indication as to which tool or strategy students should use to solve the task. Instead, she simply observes the students as they begin making sense of the task in order to decide how and when she will intervene by asking strategic questions to better understand the students' thinking. Her initial questions ("What are you guys doing?" "Is that a one?" "What does that mean in your problem?") determine the girl's understanding of place value, as represented in the standard algorithm that the student is using to solve the addition problem she has recorded on her dry-erase board. As the teacher switches to facilitate a discussion with the second partnership of students, she continues to ask assessing questions such as "Six what?," "What'd you do with that fourteen?," and "How many ones did we regroup to make a ten?" These questions help her gauge the students' ability to link their procedural and conceptual understanding of place value, as represented by the addition problem.

Engaging in formative assessment tasks, like within this short video snapshot, allows teachers to better understand students' development of mathematical thinking. The teacher uses her awareness of her students' mathematical understanding to determine her sequence of questioning in this lesson and to guide future task selection and lesson facilitation.

What Is the Evidence That the Girls Are Actively Engaged in Learning Mathematics?

In this video, students solve an addition task using either base ten blocks or the dry-erase board. We can see that the students are using different approaches to solve the problem (representing the problem using base ten blocks and solving the addition problem using the standard algorithm for multidigit addition). The students' responses to the problem-specific questions the teacher poses and her prompts for them to talk to their partners about their thinking offer evidence that students are engaged in learning about multidigit addition.

How Does the Teacher Challenge the Girls to Think Critically About the Mathematics?

We see the teacher use this formative assessment task as an opportunity to find out whether or not students are able to link their procedures for solving multidigit addition with their conceptual knowledge about place value and regrouping. The following are her specific series of questions based on the second group of partner work. As you read the questions the teacher asks and the students' responses to each, you can see the challenges to students' thinking about place value, addition, and regrouping, and the probes that the teacher intentionally uses to elicit further explanation from the students in the group.

Teacher asks, "What are you doing?"

Boy partner says, "So, we already solved the first one."

Teacher says, "How?"

Boy partner says, "So there were six here (pointing to the manipulatives), and there were eight here . . ."

Teacher says, "Six what?"

Boy partner says, "So there were 6 ones and 8 ones, and then we added them together and it was fourteen . . ."

The teacher interrupts the student to ask, "So hold on a minute, so what happened, Angela, what did you do with that fourteen?"

Girl partner says, "I regrouped?"

Teacher says, "You regrouped what?"

Girl partner answers hesitantly, "The ones . . ."

Teacher says, "How many ones did we regroup to make a ten?"

Both students remain silent, prompting the teacher to ask, "So, is this your 14 ones?"

Boy and girl partner both say, "Yeah."

Teacher goes on to ask, "So, here's your 14 ones." (She models with manipulatives.) "How many of them are there?"

Girl points and says, "Here?"

Teacher says, "Mm-hmm."

Girl partner counts aloud to ten.

Teacher says, "There's ten. Is there any way I can trade that with something else?"

The girl holds up rod.

Teacher says, "One of those? What do we call this?"

Boy partner says, "Tens."

Teacher says, "So, where do I have this 1 ten recorded here (pointing to students' dry-erase board where the students solved the problem using the standard algorithm)?"

By paying careful attention to teacher's eyes and the way she shifts her questions from one student to the other, you can also see her intentional moves to engage the girl student who is initially not contributing to the explanation of solving the problem.

You can use figure 3.8 to discuss formative assessment's role in your teaching.

How have formative assessments enriched your students' learning opportunities and your teaching strategies of mathematics?

Figure 3.8: Reflection on the role of formative assessment in mathematics instruction.

Considering Context for Tasks

Imagine being a young girl participating in a mathematics lesson. During the lesson, the task examples consist of references to football, video games, and race cars. If these activities interest you, there is a possibility you will find the mathematics exciting and engaging. If you have not engaged in these activities, are unfamiliar with these activities, or both, you have two tasks to accomplish in order to solve the posed problems you may not like. First, you need to make sense of the context of the problem-solving scenario. Then, you would need to make sense of the mathematical operations involved with solving the problem-solving scenario. While students who can relate to the problem-solving scenario have one barrier to success, the students who are unable to relate have two barriers to success. At the onset, who has a better chance at being successful in engaging in the problem-solving task? Doug Clarke and Anne Roche (2018) suggest:

> The context [of a contextualized task] serves the twin purposes of showing how mathematics is used to solve problems in the world while at the same time motivating students to solve the task. It is important in the use of contextualized tasks that the students learn something about the context as well as learning relevant to mathematics. (p. 98)

Contextualized tasks also support students as they learn how to model with mathematics (Mathematical Practice 4; NGA & CCSSO, 2010). Consider how the teacher uses a context for the task to engage all learners in this video clip of a fourth-grade class interpreting remainder, and respond to the guiding questions in figure 3.3 (page 35).

Interpreting the Remainder in Word Problems:
www.SolutionTree.com/Interpreting_the _Remainder_in_Word_Problems

Now that you had the opportunity to observe a fourth-grade lesson where students have an opportunity to write their own word problems, consider how the teacher carefully uses the role of context in task selection and lesson facilitation. We now provide additional insight via our reactions to the guiding questions.

What Actions Does the Teacher Take to Engage the Girls in Learning Mathematics?

In this lesson, the teacher plans a task in which students create their own word problems to interpret the remainder. As the teacher introduces the task, it becomes evident that many students struggle to write a word problem in which the context matches the mathematics in the problem. It seems evident that the teacher anticipates that students may struggle to make sense of the task initially as she offers follow-up questions to help the students make sense of the task and consider what types of contexts might fit the task as it is presented. Her responses include the following.

- "You can't find a number that you multiply by 4 to get 26? So, what does that mean?"
- "But I challenged you to come up with a word problem where the answer would be seven . . . that's pretty tough. Talk about that at your tables."
- "So, what's your question going to be?"
- "But, what was our goal for this task?"

Students are creating word problems to understand the mathematics. By allowing students to create their own contexts for the mathematics presented, students can select a context that is relevant to them while also making sense of the mathematics.

What Is the Evidence That the Girls Are Actively Engaged in Learning Mathematics?

In this lesson, the girls are trying to identify a context that will account for a remainder in the task. As students work to come up with their own word problems, they talk about their attempts with each other in small groups and the whole group (through specific examples shared aloud). There is evidence through students' responses, based on both the word problems written and their discussions about the sample word problems shared, that the students have difficulty creating a context that makes sense. Rather than offer an example or sample context, the teacher asks facilitative questions to help the students make sense of the remainder and how to represent it in the individual contexts.

How Does the Teacher Challenge the Girls to Think Critically About the Mathematics?

This task calls on students to create their own scenarios through writing word problems. The challenge for most learners is based on the fact that the scenarios that they need to create have to make use of a remainder in a division problem. For many, this involves thinking beyond the replication of other division word problems, with which they seem familiar. The teacher challenges her students to think critically about whether or not their contexts meet the challenge of the task (in which the solution to the word problem would be seven, and not six). Although there is evidence of a challenge for most students, the teacher recognizes the value of the productive struggle this rich task presents. The task allows students to think more critically and successfully write their own word problems representing the mathematics at hand.

You can use figure 3.9 to determine if any word problems or books in your instruction have a gender bias.

 Look back at the word problems you have given your students or the books you have read to your students. Are the word problems or books geared to a certain gender? Are there typical stereotypes hidden within the context?

Figure 3.9: Review of teaching material for gender bias.

Modeling of Mathematical Power

Of course, you know that teachers play a key role in students' mathematics learning experiences and the quality of students' overall schooling experience (Schmidt, Burroughs, Cogan, & Houang, 2017). It is the teacher that has the primary relationship with students in the classroom. As a teacher, you also have the primary role of setting the atmosphere of the environment; making decisions about curriculum, instruction, and assessment; and deciding how you want students to interact in the learning community.

You are a role model for students. What kind of role model you will be and how you will model mathematical power are up to you. Students particularly need to learn what it means to do mathematics and how one should behave mathematically. When you model mathematical power for students, you show students a preview of their own possibilities as learners of mathematics. That is powerful!

Modeling mathematical power can have a direct impact on girls' (and all learners') confidence and self-perception as a learner of mathematics; this connects to Mathematical Practices 1 and 3, in particular as it relates to girls persevering, problem solving, and interacting with others in the mathematics classroom (NGA & CCSSO, 2010). Consider how the teacher models mathematical power and promotes increased engagement and confidence among girls in the video clip of a fourth-grade class using benchmark fractions, and respond to the guiding questions in figure 3.3 (page 35) as you watch.

 Comparing Fractions Using a Benchmark:
www.SolutionTree.com/Comparing_Fractions_Using_a_Benchmark

Now that you had the opportunity to observe this fourth-grade lesson on using benchmark fractions, consider how the teacher models mathematical power through her task selection and lesson facilitation. We now provide additional insight via our reactions to the guiding questions.

What Actions Does the Teacher Take to Engage the Girls in Learning Mathematics?

In the video, you see the teacher present the task and facilitate students' responses to encourage a variety of individual ideas about how the fractions $3/8$ and $5/8$ compare to one another and the benchmark fraction of ½. The teacher asks students to explain their thinking and offer additional examples to support their responses (for example, "Can you explain your thinking? What do you mean by model?"). Throughout the lesson, the teacher remains fully present and engaged in students' explanations and justifications, asking content-specific questions related to the ideas that they are sharing with their classmates. In this way, the teacher is not only engaging the students in learning mathematics but also demonstrating her own confidence in the mathematics content knowledge that is necessary in order to solidify students' awareness of using benchmark fractions as a strategy to compare fractions.

What Is the Evidence That the Girls Are Actively Engaged in Learning Mathematics?

The girls who contribute verbally in this lesson are selected both by their own willingness and also the teachers' prompts for them to share their thinking. Throughout the lesson, all students engage in learning about mathematics as they respond to the teachers' sequence of questions about comparing the fractions presented. The teacher highlights students effectively using reasoning skills in the way they compare the fractions.

How Does the Teacher Challenge the Girls to Think Critically About the Mathematics?

By presenting another student's strategy to compare the fractions in this task (a scenario in which a student compares ⅜ and ⅝ in relation to the benchmark fraction of ½), the teacher asks students to think critically about that approach and to make sense of it through their own explanations.

The teacher also plans for a series of tasks in which students compare a set of fractions with like denominators and then compare a set of fractions with like numerators but unlike denominators. Presenting the tasks in this sequence encourages students to think more about the possible strategy of using benchmark fractions (in this case, the benchmark of one) because for some, their original strategy to compare fractions with like denominators would no longer be helpful with fractions with unlike denominators.

You can use figure 3.10 to brainstorm ways to engage girls and build their confidence in mathematics.

 How do you model your mathematical power to promote increased engagement and confidence among girls in your classroom?

Figure 3.10: Ways to promote engagement and confidence for girls learning mathematics.

Conveying Teacher Expectations

A big part of how students perform in the classroom is based on teacher expectations. According to M. Katherine Gavin and Sally M. Reis (2003) of the Neag School of Education at University of Connecticut:

> One of the main reasons that girls do not succeed in mathematics may not be due to any lack of ability or effort—but rather it may [be] attributed to the fact that they are not expected to excel in this area by some of their parents, teachers and peers. (p. 33)

Gavin and Reis (2003) further claim that "evidence also exists that girls are regarded as less capable in mathematics by some of their teachers and parents, and these perceptions may influence girls' opinions of their own abilities" (p. 34). To combat this issue, NCTM's (2016) position is that teachers should have high expectations of students and manifest these high expectations in at least five ways.

1. Acknowledge the mathematical ability of each and every student.
2. Help each student develop a positive perspective about his or her relationship with mathematics.
3. Create learning opportunities that make use of students' knowledge and experiences.
4. Teach for reasoning and sense making.
5. Develop learning experiences that engage each and every student.

When girls struggle with mathematics, respond appropriately. Juli K. Dixon, Edward C. Nolan, Thomasenia Lott Adams, Lisa A. Brooks, and Tashana D. Howse (2016) suggest that "by preparing for common misunderstandings and errors, you will be better able to help students successfully engage with mathematics and overcome barriers to understanding" (p. 136). Have balanced expectations for all students. Avoid making success in mathematics as something that is gender-centric. One way educators can do this is by conveying high, positive expectations to all students. Secondly, educators can make sure that all students have access to high-quality mathematics (NCTM, 2016).

Teachers should make sure high expectations for girls (and all students) are explicit. Consider how the teacher makes high expectations explicit in the video clip of a third-grade small group working on area problems, and respond to the guiding questions in figure 3.3 (page 35) as you watch.

Find Area and Order by Size:
www.SolutionTree.com/GR3AreaOrder

Now that you have the opportunity to observe a third-grade, small-group lesson on area, consider how the teacher demonstrates her high expectations for all students in the group through her task selection and facilitation of the lesson. We now provide additional insight via our reactions to the guiding questions.

What Actions Does the Teacher Take to Engage the Girls in Learning Mathematics?

In the video, the teacher gives each student the opportunity to represent the task using his or her own strategy. This teaching decision gives everyone access to the problem. This is an example of drawing on each student's background knowledge and acknowledging each's mathematics ability. The teacher uses verbal and nonverbal ways to demonstrate her expectations, which the following examples show.

- She provides all students with time to engage in the task and share their initial strategies, and then additional time to re-engage in the task and work toward a solution. This conveys the expectation that all students will work toward a solution to the task, not just those students who may finish first.

- When a question arises from a student in the group, the teacher explicitly communicates her expectation that all students need to understand the question and be prepared to support each other by stating (to the girl student with a question), "Show us what you would write. I want us to make sure that we understand her question and see if we can help her."

- When the teacher identifies that students disagree about the accurate way to calculate the area of the shape, she conveys her expectation that students need to share their thinking with one another in order to make sense of the task by stating, "I hear yes and no. It sounds like you guys have some discussion to have."

What Is the Evidence That the Girls Are Actively Engaged in Learning Mathematics?

As the lesson continues, you see two student partnerships begin to represent the area of a rectangle in different ways using the square tiles available for the task. The teacher realizes (and even verbalizes) that the students are using the tiles in different ways to determine the area of the rectangle on the paper in front of them, yet she does not try to change the approach that either pair is using (Mathematical Practice 5; NGA & CCSSO, 2010). Instead, she uses their diverse representations as an opportunity to validate their own strategy use and promote student discourse about the mathematics that they are each representing. While this teacher move has commonalities with the first possibility presented in this chapter, honoring diverse ways of doing mathematics, it also provides all students the time they need to come to a solution themselves with the knowledge that they will be able to do so. She values each student's approach, which likely contributes to the students' high levels of engagement and positivity as they engage in the task.

How Does the Teacher Challenge the Girls to Think Critically About the Mathematics?

Notice that the opportunity to have one's own strategy valued may contribute to an environment where students are willing to take risks and be vulnerable as they share their thinking. For example, in this instruction, one girl asks what she should do next to calculate the area of the rectangle based on the way that she is choosing to represent and calculate the area with the square tiles. The teacher recognizes this opportunity to teach reasoning and sense making, even beyond the content that links to the task (Mathematical Practices 3, 6, and 8; NGA & CCSSO, 2010). You might imagine that if the teacher had demonstrated a more rigid expectation that all students represent area in the same way for this task or if she had been more focused on the solution than the students' strategy use, engagement, and discourse, then the girl who was willing to verbalize her question may not have been as forthcoming with sharing.

You can use figure 3.11 (page 52) to reflect on strategies you use to foster an inviting environment where students are comfortable taking risks.

 What are some strategies that you use in your classroom or school to promote an inviting and safe environment where students are willing to take risks to better understand the material?

Figure 3.11: Reflection on classroom strategies that foster a safe environment for risk taking.

Planning for Positive Practices for Girls Studying Mathematics

You are the driving force behind creating a classroom environment that can empower girls' awareness of mathematics. Knowing who you are teaching and how you will teach them will predict the future mathematics success for the girls in your class. How do educators create that environment? We answer the question in three contexts: (1) the classroom, (2) the school and district, and (3) the school-home connection.

The Classroom

As noted previously, for each of the eight possibilities we have created an instructional tool to help teachers put the possibilities into practice.

To enhance instruction that honors diverse ways that girls do mathematics, try four additional teaching strategies (see figure 3.12).

1. Provide a task and instruct students to solve it in multiple ways. When students can apply different strategies to solve a problem, it forces them to be more flexible with their thinking. It provides them with an opportunity to compare and contrast different ways of knowing and doing mathematics.

2. Provide tasks that have more than one acceptable answer. Doing this allows much more opportunity for students to talk and engage in mathematical discourse by justifying and proving their answers. Students will spend less time worrying if their answers are correct and more time on how to solve the problem.

3. Give students an opportunity to share their problem-solving strategies within small groups and the whole class. Part of doing mathematics is not only being able to solve a problem but also being able to explain a strategy using precise vocabulary and make sense of other students' thinking. Create a culture in the classroom that empowers students to learn from one another by listening as well as sharing their strategies and to know how to ask clarifying questions in order to seek understanding.

4. Instruct students to discuss the pros and cons of different strategies for the same task. With every problem comes multiple strategies to find the solution. The end goal is for students to have a variety of problem-solving strategies and know which strategy is most efficient for the problem they are solving.

Figure 3.12: Four additional teaching strategies for honoring diverse ways of doing mathematics.

Visit go.SolutionTree.com/mathematics for a free reproducible version of this figure.

To enhance instruction that promotes classroom discourse, consider using the teaching tool *discourse cards*. When organizing students to solve problems and engage in discussion around their mathematical thinking, discourse cards can support students to build language that strengthens their communication skills. The discussion prompts and sentence starters in the discourse cards enable students to practice respectfully and constructively agreeing and disagreeing with one another as they share their mathematical thinking. Students are also supported in making their thinking visible by using the prompt their teacher provides on a selected discourse card to extend discussion. See figure 3.13 for examples of questions and sentence starters to prompt discussion in the discourse cards.

The following are questions that promote discussion.
- What strategy did you use to get your answer?
- How do you know your answer is correct?
- Can you solve your problem in another way?
- How did you know which operation to use?
- Do you have anything to add?
- How do you know your answer is right?
- Did anyone get the same answer a different way?
- How did you know which operation to use?
- What if _____?

The following are sentence starters that promote discussion.
- I agree with you because _____.
- I disagree with you because _____.
- What I heard you say was _____.
- I see it another way. Let me explain.
- The evidence I have to support that is _____.
- The strategies I used were _____.
- Explain your partner's strategy.

Figure 3.13: Sample discourse card questions and sentence starters.

*Visit **go.SolutionTree.com/mathematics** for a free reproducible version of this figure.*

To enhance instruction for girls to engage in hands-on learning, consider the following teaching practice: create a mathematics toolbox (see figure 3.14).

Create a mathematics toolbox with a variety of manipulatives and tools your students can have easy access to as they solve mathematical tasks. You can create one per student or group. Put all tools in a plastic container. Your mathematics toolbox might include two-sided counters, base ten blocks, hundreds chart, inch square tiles, ten frames, Unifix Cubes, ruler, centimeter cubes, and geometric shapes. Students can use these tools to make connections between multiple representations to enhance students' understanding of concepts.

Figure 3.14: Creating a mathematics toolbox.

*Visit **go.SolutionTree.com/mathematics** for a free reproducible version of this figure.*

For the possibility of using questioning to boost understanding, we have included two tools, one for assessing questions and one for advancing questions. *Assessing questions* uncover students' thinking and understanding in the moment and provide an opportunity for the teacher to use students' responses on how to move forward in instruction. They ask students for an explanation, to link between concepts, to probe their thoughts, and to consider the strategies in use. See figure 3.15 for examples of outcomes that result from assessing questions. Notice the different types of questions a teacher might use to engage students.

Seeking Examples
How could your strategy work with another problem?
How does this relate to _____?
Seeking Nonexamples
Would this work every time?
Can you think of any examples that don't work?
Requesting Explanations
Can you convince me of your solution?
What did this calculation give you?
Linking Between Concepts
What is happening in these two situations? How are they similar? How are they different?
What is the relationship between _____ and _____?
Probing Thoughts
What would happen if _____?
Can you think of a counterexample?
Soliciting Strategies
What steps are involved?
What mathematics vocabulary did you use or learn?

Figure 3.15: Sample assessing questions.

*Visit **go.SolutionTree.com/mathematics** for a free reproducible version of this figure.*

Teachers use *advancing questions* to move students forward in their thinking and prompt students to transfer their understanding to new and different situations (Larson et al., 2012). They ask students for clarification, to consider different cases, to avoid generalizations, to consider alternative explanations, and to apply the content to real life. See figure 3.16 for examples of outcomes that result from advancing questions.

Seeking Clarification
What does this part represent in your solution?
What do you see as other possible methods of solving this problem?
Connecting to Different Cases
What other mathematics can you connect this with?
Where else have we used this?
Making Generalizations
How do you know your answer is reasonable?
What did you discover while solving this problem?
Finding Alternate Explanations
Would this work with other numbers? Can you think of a counterexample?
How can you solve this a different way, but arrive at the same answer?
Applying Content
Can I _____?
Is there a real-life situation where you (or someone else) can use this?

Figure 3.16: Sample advancing questions.

*Visit **go.SolutionTree.com/mathematics** for a free reproducible version of this figure.*

Formative assessment is a powerful way for teachers to gauge students' progress as they learn and make decisions to support them wherever they are. For this possibility, we have included a formative assessment observation tool to help you break this process down (figure 3.17, page 56).

As you are considering the context for tasks in relation to your own classroom, let's begin by reflecting on the girls you are teaching (figure 3.18, page 56).

As part of creating a safe classroom environment, knowing who you are teaching and establishing positive relationships with your students are parts of creating a safe classroom environment (Yoder, 2014). If answering these questions came as a challenge, the following teaching tool will be especially useful for you. Knowing your students will guide you into creating meaningful lessons with your students based on their interests.

During the first week of school, spend some time doing get-to-know-you activities to better connect with your students throughout the year. One way to discover your students' interests is to have them fill out an interest survey. See an example of an interest survey in figure 3.19 (page 57).

Date:

Formative Assessment Observation

Learning Target:
Learning Task:
Questions to Elicit Understanding:

Based on observations from the task:

What do my students know?	
What are my students able to do?	
What strategies are my students using?	
What struggles are my students having?	
What are my next teaching steps tomorrow?	

Figure 3.17: Formative assessment observation tool.

*Visit **go.SolutionTree.com/mathematics** for a free reproducible version of this figure.*

Answer the following questions about the girls in your classroom.

1. What are they good at in school?
2. Do they prefer to work alone or in small groups?
3. Have they had any positive or negative experiences in mathematics?
4. What hobbies or extracurricular activities do they participate in?
5. What do they want to be when they grow up?

Figure 3.18: Teacher reflection on girls they teach.

*Visit **go.SolutionTree.com/mathematics** for a free reproducible version of this figure.*

> Answer the following eleven questions about your interests.
> 1. What games do you like to play?
> 2. What music do you like to listen to?
> 3. What is your favorite hobby?
> 4. What kinds of books do you like to read?
> 5. Do you have a pet? If so, what do you have, and what is your pet's name?
> 6. What is your favorite food?
> 7. What is your favorite movie?
> 8. What is your favorite holiday?
> 9. What is something that really matters to you?
> 10. What makes you laugh?
> 11. What do you like to do on the weekends? What about in the summer when school is out?

Figure 3.19: Sample student interest survey.

*Visit **go.SolutionTree.com/mathematics** for a free reproducible version of this figure.*

Using the results of the interest survey to write your word problems or build mathematical tasks around their hobbies or favorite foods is one sure way to engage your students. If you find that the girls in your class are highly interested in gardening, create mathematical tasks revolving around gardening. The following are sample grade-level tasks you might consider for this interest.

- **Kindergarten:** Aubrey picked 10 flowers from her garden. Some flowers are pink, and some flowers are yellow. How many flowers are pink? How many flowers are yellow?

- **Grade 1:** Your class is going to plant a vegetable garden with 2 different vegetables. In order to select which vegetables to plant, survey other first-grade students about which vegetable is their favorite: green beans, spinach, peas, or carrots. Then, create a graph to display these data. Based on the data you collect, which 2 vegetables will you plant?

- **Grade 2:** Farmer Beth was going to plant some carrots in rows. She wants to plant 16 carrot plants all together. Each row has to have an equal amount. How many carrots does she need to plant in each row? Explain your thinking to a partner. Can you think of another way?

- **Grade 3:** You have been asked to design a flower bed for a spring garden for tulips, marigolds, and daffodils. The flower bed must have 3 adjoining rectangles of different sizes and have a total area of 36 square feet. Draw your design for the flower bed. Compare your design to a partner. How are they similar? How are they different?

- **Grade 4:** A florist went out to her rose garden to cut flowers for an upcoming wedding. She needs to create a vase arrangement with the roses for 8 tables. She cut 58 roses to put in vases. Help the florist design her vase arrangements by putting an equal amount in each vase. How many roses will each vase have? Will she use all the roses?

- **Grade 5:** Sonja and her brother are weeding their garden. Sonja collects 4 times as many bags of weeds as her brother. If her brother collects $\frac{1}{6}$ of a bag of weeds, how many bags of weeds does Sonja pick?

Now you have incorporated data from the student-interest survey to engage learners. Do you know any local women gardeners? This would be an ideal time to invite them into the class to share their expertise with your students and let them share how they use mathematics on a daily basis as gardeners. This builds positive relationships between teacher and student and student to student before learning can take place. Keep in mind that this is an example of how you can use an interest survey to engage the girls in the class. You can do the same with the responses from the boys in the classroom. It's important to keep a balance between the boys and girls and be aware of the type of problems you are providing.

Shifting now to the possibility of modeling mathematical power, we have created a simple activity that teachers can do together with students to help them build and sustain their own self-confidence (figure 3.20).

Our tool for the eighth possibility, conveying teacher expectations, is a self-reflection chart that you can use to be mindful of how your classroom environment affects students and how you can use it to convey what you expect of them (see figure 3.21).

Any of these tools can be useful to any teacher, and we expect you to adapt them as necessary as you weave our ideas into your own classroom. Just as there are many ways to solve a problem, there are many ways to ensure that girls are engaging and succeeding in your classroom.

One third-grade teacher, Toni Laughrey, shares her strategy for teaching girls in mathematics:

> As a math teacher, I love engaging with my students and hearing their thoughts and ideas on various strategies. In creating a "safe space" where it's okay to be wrong and take risks with solving problems, I have seen so much more confidence among my girls. They are excited to try new things and share these thoughts with each other, and when they are incorrect, they are eager to find out why and learn new techniques to use. I not only want them to succeed for the obvious reasons, but for their own confidence and realization that they have so much knowledge within them, it's all about unlocking it and discovering how to use it. (T. Laughrey, personal communication, November 1, 2017)

Were you able to hear the voice of this teacher and pick up on her passion as she engages her girls in mathematics? Once you know who you are teaching, it is time to nurture an environment for doing mathematics. A place where your students can collaborate to problem solve and think mathematically with their peers. Somewhere they can ask questions, persevere, and have successful experiences problem

Modeling Mathematical Power

At the end of a mathematical task, have each student in the group pick a question to answer. Consider sharing your responses to students as well, both from your experiences as a teacher and a learner of mathematics.

- I felt confident when . . .
- I felt proud when . . .
- I was able to help others when . . .
- I was a leader when . . .

Figure 3.20: Tool for modeling of mathematical power.

Visit go.SolutionTree.com/mathematics for a free reproducible version of this figure.

What Does Your Classroom Say About You and How You Convey Your Expectations of Your Students?		
Reflection Questions	**My Current Classroom**	**Considerations or Changes I Might Make**
Does your seating arrangement offer students opportunities to collaborate? How do you decide who sits where?		
What are your expectations for student engagement? Where did they come from? Do all your students know what they are?		
What does your classroom sound like during a task? Look around and take note of who is talking and who is quiet.		
How do you empower your students? What is your response when a student gives a wrong answer? How do the other students react?		
Why would a student want to be in your class? What motivation does a student have to come back to school tomorrow?		

Figure 3.21: Teacher self-reflection tool for conveying expectations.

*Visit **go.SolutionTree.com/mathematics** for a free reproducible version of this figure.*

solving because they engage in *doing* mathematics and not just showing how to solve problems. Think back to the last mathematics lesson you taught from beginning to end. What was your role? What were the students doing? Who was doing the sense making? The classroom video clips in this chapter highlight that many teaching practices through teacher's facilitation move in whole-group, small-group, and one-on-one discussions. If the girls in your class are going to become mathematically proficient, they need the environment to do so.

Your teaching practices directly impact your students' success. The girls in the class must see that you treat them no differently than the boys and expect no less from them than the boys. Through encouragement and targeted feedback, girls need to realize they are just as good at mathematics as boys, as the research suggests. Provide extra opportunities for girls to build on areas of weakness and help them develop positive self-perceptions. Be sure to call on them just as often, and carefully monitor that the depth or level of your questioning does not change based on student gender. A good way to track this is to invite someone into your classroom to observe your questioning or to record yourself to later reflect on your teaching.

In order to impact girls' learning of mathematics, we need to know who we are teaching, and we need to communicate high expectations for *all* students by incorporating culturally relevant pedagogy. Not all girls will exhibit mathematics anxiety, but there are many that do, and your goal is to promote confidence in their mathematical ability. You can adjust your current teaching practices to prepare the girls in your classroom by doing the following.

- Provide more emphasis on understanding and application of problem solving than on the answer to a problem to show that you expect girls to demonstrate thinking and reasoning.

- Offer opportunities to build spatial awareness and develop visualizing for students, paying particular attention to girls in your class who struggle with spatial skills. Involve them in building, designing, and solving puzzles. (See chapter 1, page 7, for more ideas to build spatial awareness.)

- Foster self-confidence in girls by giving them immediate feedback when they succeed. It is important for them to feel validated that they are good at what they are doing. One way to provide ongoing feedback is to have regular one-on-one conferences with students and provide specific and personalized feedback on the work they are doing. Be descriptive as you highlight the work they have done. At times you will also need to give feedback when a student has failed or been unsuccessful. You can approach this by beginning with something positive they have done. Then provide actionable feedback about what they can do differently, and end with encouraging words. Consistent feedback is a powerful way to influence academic achievement.

- Use team-building activities early in the year to foster an environment where students feel comfortable to ask questions, make mistakes, and take risks. This classroom culture helps students better understand and learn more about each other. One great beginning-of-the-year activity for the whole group is called "Find Someone Who . . . ." Students go around and identify one of their peers based on a certain listed attribute. For example, "Find someone who has a pet." Another one is having small groups compete to complete a task, such as building the tallest tower out of uncooked spaghetti and marshmallows. Any activity giving a team one goal to accomplish and requiring the members to collaborate to reach that goal challenges students to problem solve and work well with others.

The School and District

Schools and districts can wield great power to ensure improved, sustained learning for every student. The best way we've found to achieve this goal is the professional learning community (PLC) process. Teachers and staff members at schools that adopt the PLC process collaborate and plan for positive practices through discourse about the mathematics curriculum and students' mathematical learning. That's because a PLC is a whole-school or whole-district effort in which "educators work collaboratively in recurring cycles of action research to achieve better results for the students they serve. PLCs operate under the assumption that the key to improved learning for students is continuous job-embedded learning for educators" (DuFour, DuFour, Eaker, Many, & Mattos, 2016, p. 10). There are three big ideas that drive the work of a PLC.

1. **A focus on learning:** This means PLC schools ensure that all students learn at high levels. It's not acceptable that all girls are not excelling at high levels in their mathematics studies.

2. **A collaborative culture and collective responsibility:** Teachers work together. There's no such thing as *my students* and *your students*. In terms of mathematics teaching and learning, every teacher, coach, and administrator is responsible for girls' success.

3. **A results orientation:** A PLC focuses on results—evidence of student learning. Teachers, coaches, and administrators assess the effectiveness of their practices for teaching girls mathematics and use the results of that assessment to improve their professional practices and respond to individual girls who need intervention or enrichment.

Reflect on these big ideas in regard to your mathematics team and the focal points of discussion. Would you consider your team to impact change? If the school primarily emphasizes standards and data only, it is time it shifts the conversations around how your teaching practices can influence student achievement.

Consider a collaborative team of third-grade teachers who are planning for an upcoming unit on fractions. The group took the time to unpack the standard and make sense of what the students had to learn by creating learning goals. Its next step was looking through its textbooks to see which lessons teachers were going to teach on each day of the unit and using additional resources to create mathematics centers.

Although collaboration was taking place around a standard, the teachers didn't select or develop quality tasks to differentiate their instruction to meet the different learners within their classrooms' needs. This would have been an opportune time to address not only achievement levels but also gender differences to engage all learners.

Preparing for an upcoming unit takes careful and thoughtful planning within a PLC. Selecting tasks is an important step toward planning for a lesson, but it is also a critical opportunity to consider the teacher's role in the lesson. Anticipating how students will solve the task, creating questions to facilitate their thinking, and planning for your next teaching steps all are pieces to leading your students to make sense of mathematics. Collaborative conversations before and after a lesson or unit are important aspects of the teaching and learning process. Using the tasks, question, and evidence (TQE) process (see chapter 4, page 67) that Juli K. Dixon, Edward Nolan, and Thomasenia Lott Adams (2016) created offers a concrete way to engage in this type of intentional planning. In this process, teachers select a task that aligns to content standards that will engage students in mathematical concepts, develop questions to facilitate during the task, and plan what to do with the evidence they collect to further deepen students' understanding of mathematics.

Possibilities about girls as learners of mathematics do not just exist at home or in the classroom. District and school leaders' voices are the foundation to influence and educate teachers and parents to establish best practices within the classroom and home. Teacher collaboration with administration and instructional coaches is essential to promote positive mathematics experiences with the emphasis of selecting content to engage and challenge girls.

Regardless if a school or district is a PLC, leaders can positively impact the future success of girls in mathematics. Through careful planning and consideration, there are four ways that impact can occur.

1. Provide mathematics professional development for teachers to build their content knowledge. It is common that for each school year, teachers move to different grade levels and have to learn a whole new set of standards. Professional development with a focus on teachers being able to engage in hands-on learning experiences to support their own understanding and clarify any misconceptions will enable them to teach more confidently in the classroom.

2. As a district or school, examine the schedule to bring mathematics to the forefront by providing blocks of time just for problem solving. By balancing the time students engage in reading and mathematics, teachers will see you value mathematics as well.

3. Provide meaningful tasks within the district mathematics instructional plan that align to the standards that teachers can access when planning in their collaborative teams.

4. Arrange for teachers to have a mathematics coach available to support them with best teaching practices and planning meaningful mathematics lessons for girls.

The School-Home Connection

Home serves as the first learning environment for students. As a teacher, making a school-home connection will serve as another key piece to building girls' confidence in mathematics. Not only is it essential to build relationships with your students, it is just as important to connect with your students' families. Starting early in the school year, plan to find occasions when you can build rapport by joining parents in the teaching of their daughters. You might consider trying to:

- Reach out to parents with welcome phone calls, emails, or letters (or even home visits if warranted) prior to or at the start of the year
- Plan for engaging low-risk mathematics tasks (such as tangram puzzles) during meet-the-teacher or open house events
- Send home manipulatives that families can use throughout the year to engage in mathematics tasks together (for example, pattern blocks, tangrams, or two-colored counters)

Sometimes, we do not realize how much families influence their children. A third-grade teacher, Jordyn Maher, recalls her early mathematics experiences as a child and how her father impacted her:

> As a young girl, I remember coming home from school feeling frustrated and saying, "I hate math! It's too hard!" My dad sat me down and began telling me how much he loved math. Hearing the way he talked about it made ME want to love math just like him. This motivated me to take more of an interest in it and push myself to understand what I was being exposed to. I share this with my students each year in hopes that my personal experience will inspire them to look past the "this is too hard!" and be able to see the fun in math! (J. Maher, personal communication, November 1, 2017)

Teachers teach students. Teachers teach parents. To guide your students to their fullest mathematical potential, you will need to educate and inform parents of ways to empower their daughters' mathematical ability at home. For example, you might write a newsletter that features columns to do just that, such as the sample in figure 3.22.

Parent Connection

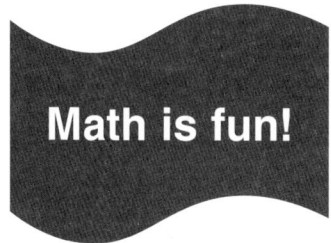

Ways to Empower Your Daughter Mathematically at Home

Parents are important partners to ensure mathematical success for their daughters. By engaging your daughter in interesting and fun mathematical activities, you are building a foundation for a lifelong enjoyment of mathematics. It is important to foster a positive attitude toward mathematics at home by letting your child know you believe in her abilities. As a parent, you are a significant contributing factor to your daughter's success. Here are some fun ways to bring mathematics into your everyday life. Please feel free to call with any questions you ever have.

Homework Help

Helping your child with her mathematics homework can be challenging at times! Here are some guiding questions that are designed to help your child think through her mathematics homework problems. If she gets stuck, ask the following questions.

- What do you need to figure out?
- Did you solve a problem like this one in class today?
- What have you tried so far?
- Does your answer make sense?
- Can you make a drawing to help you think about the problem?
- Can you explain in your own words what the problem is asking?

Math at Home

- At the grocery store, ask questions that require your daughter to add, subtract, multiply, or divide.
- Follow a recipe and bake together. Create a challenge by doubling or halving the recipe.
- Plan a family event or trip together. Figure out how much money you need to cover all costs.
- Have a family game night. Play games that encourage spatial sense or strategies to win.
- Plan a garden outside. Use measurement to space out plants.

Figure 3.22: Sample newsletter to inform parents how to support mathematics education at home.

Keep in mind, we all had different mathematics learning experiences growing up. Most families want to help their children learn and are willing to do what they can to help. In some cases, however, an obstacle that interferes with the support they provide is not understanding the mathematics in general, as evidenced by the following quotes from parents of elementary students.

- "First, I have to see the problem and find an example to understand how it is done, then I can attempt to explain how it is done." (F. Dailey, personal communication, December 3, 2017)

- "Not knowing new terminology for things like turn around facts and make a ten and not having math tools for them to use at home is the biggest challenge for me." (A. Jordan, personal communication, December 3, 2017)
- "I have a third grader and not having resources for parents to understand the math or examples to try and help my daughter is a challenge." (A. Richter, personal communication, December 3, 2017)
- "The language. Growing up, fractions greater than one were called improper fractions. If you are not immersed on a daily basis it's hard to keep up." (E. Matrani-Morse, personal communication, November 27, 2018)

Building awareness of and relationships with family and caregivers of students is just as crucial as the teaching practices in the classroom that can positively impact the girls we teach. Finding ways to inform parents about how they can help their daughters will impact how the girls in your class succeed. Consider three practices to increase the impact the home environment can have on your students.

1. Plan how you will build school-home relationships to meet *all* families and how you will reach them. Find out what needs the families have and what you will communicate to them.
    - Send out a parent survey at the beginning of the school year with a variety of ways to communicate, such as a phone call, email, text, conference, or note, to determine the best way to communicate.
    - Invite families in before school to get to know them as well as meet other families in the class. You could host a "Data and Donuts" event and let the students share their data or work they have done in class.
2. Establish homework your students can have success with and complete independently to avoid struggles at home that can lead to mathematics anxiety.
    - Send home homework that reviews and allows students to practice what they have learned rather than new content they have yet to master.
    - Be creative with homework and have students create a board game, write word problems, or play a game with a family member.
3. Collaborate with your team, school, or district to create mathematics family nights, which serve as a positive mathematics experience for families to engage in.
    - Invite families to come and play mathematics games together and experience mathematics as something that can be fun.
    - Offer parents opportunities to participate in mathematics nights designed to help them learn the mathematics their students are learning or provide them with strategies they can use to help their children succeed at home.

Conclusion

As you explored the eight possibilities presented through the classroom videos, teaching tools, and narratives in this chapter, you likely found yourself making connections to your own students and classroom practices. Consider the actions that you might engage in based on your takeaways from this chapter. How will you prepare and facilitate opportunities for girls to learn mathematics effectively in your classroom? Which of the tools provided might you use in the coming weeks? What examples from the classroom videos might you try to emulate in your own teaching? As you have read (and likely know yourself), the mathematics experiences that you are providing for your students in this single year of their education career are critical and will impact their future success.

Figure 3.23 offers actions to support girls learning mathematics.

> The following three actions can help you consider the possibilities in regard to teaching girls mathematics.
>
> 1. Encourage girls to ask questions. Often girls with mathematics anxiety are reluctant to ask questions and do not know what questions to ask. Create a system in your class to offer support to those reluctant students. Have a question card or question on board students can use as a guide with the appropriate questions to ask.
> 2. Encourage your students to make a list of careers that involve mathematics that interest them. Ask them to select one of the careers and make a web of the qualities you need in order to be successful in that career.
> 3. Encourage families to watch a sports game on the television with their daughter, look at graphs or charts in the newspaper, or follow a recipe to cook something for the family. Offer examples of age-appropriate mathematics questions that families might ask. (Be sure to encourage questions that stimulate explanations and justifications.) For example, Which team is winning? What is the difference in the scores? What do you notice about this graph? How much would we need if we doubled the recipe?

Figure 3.23: Actions for considering possibilities in teaching girls mathematics.

Reflections

Answer the following five questions independently or in your book study to further your understanding and goals related to teaching girls mathematics.

1. What were some similarities and differences between your teaching of mathematics and the instructional techniques of the teachers in the classroom videos in this chapter?
2. What are some key takeaways from this chapter that you might share with your colleagues or team?
3. How intentional have you been in making connections between your mathematics instruction and your students' families? How might you try to support your students' positive experiences with mathematics at home?
4. You've likely heard the phrase "possibilities are endless." While that may be true, in this chapter, a mere eight possibilities can positively impact the mathematical learning experiences of your students. What do you think? Which will you try first?
5. What examples of mathematics instruction from this chapter might you use or adapt in your own classroom? Consider the tasks that were presented and the mathematics manipulatives that were used.

Further Reading

Amelink, C. T. (2012). Female interest in mathematics. In B. Bogue & E. Cady (Eds.), *Apply Research to Practice (ARP) resources.* Accessed at www.engr.psu.edu/AWE/ARPResources.aspx on September 17, 2018.

Gojak, L. M. (2013). *Partnering with parents.* Accessed at www.nctm.org/News-and-Calendar/Messages-from-the-President/Archive/Linda-M_-Gojak/Partnering-with-Parents on August 31, 2018.

Gresalfi, M. S., & Chapman, K. (2017, April). *Recrafting manipulatives: Toward a critical analysis of gender and mathematical practice.* Paper presented at the 9th International Mathematics Education and Society Conference, Volos, Greece.

Rellensmann, J., & Schukajlow, S. (2017). Does students' interest in a mathematical problem depend on the problem's connection to reality? An analysis of students' interest and pre-service teachers' judgments of students' interest in problems with and without a connection to reality. *ZDM Mathematics Education, 49*(3), 367–378.

Soni, A., & Kumari, S. (2017). The role of parental math anxiety and math attitude in their children's math achievement. *International Journal of Science and Mathematics Education, 15*(2), 331–347.

CHAPTER 4
Priorities for Teaching Girls Mathematics

Advocate for gender equity in the STEM disciplines at the school and community levels, as well as that of the wider society. This means sharing information and strategies with colleagues, parents, and students themselves, as well as seeking formal structural changes or policies and programs that will help forward this important educational agenda.

—Lynda R. Wiest, Professor of Mathematics Education and Educational Equity at the University of Nevada, Reno

Imagine being in a place or aspiring to go to a place where no one is the same gender as you. There is no one in this place that is like you. It appears that all are the same but you. Everyone excels in mathematics, except for you. You are a single outlier. In figure 4.1, consider how that would impact you.

Respond to the following.
- How would you feel in an environment as an outlier? Would you feel uncomfortable?
- If you would feel uncomfortable, what would make you feel welcomed?
- Would you prefer that you were in a different environment?
- How would you want to interact with others in your environment?

Figure 4.1: Considering the impact of feeling like an outsider.

Visit **go.SolutionTree.com/mathematics** *for a free reproducible version of this figure.*

How might *priorities* support and strengthen girls' experiences as learners of mathematics? In order to provide the experiences, tools, and support networks for all learners, teachers need the right priorities for teaching girls mathematics. In this chapter, we highlight four key priorities that can impact the success and achievement of students in the field of mathematics, with an emphasis on empowering young girls.

1. Equity
2. Teacher beliefs
3. Opportunity
4. Teacher knowledge

Figure 4.2 (page 68) illustrates these priorities.

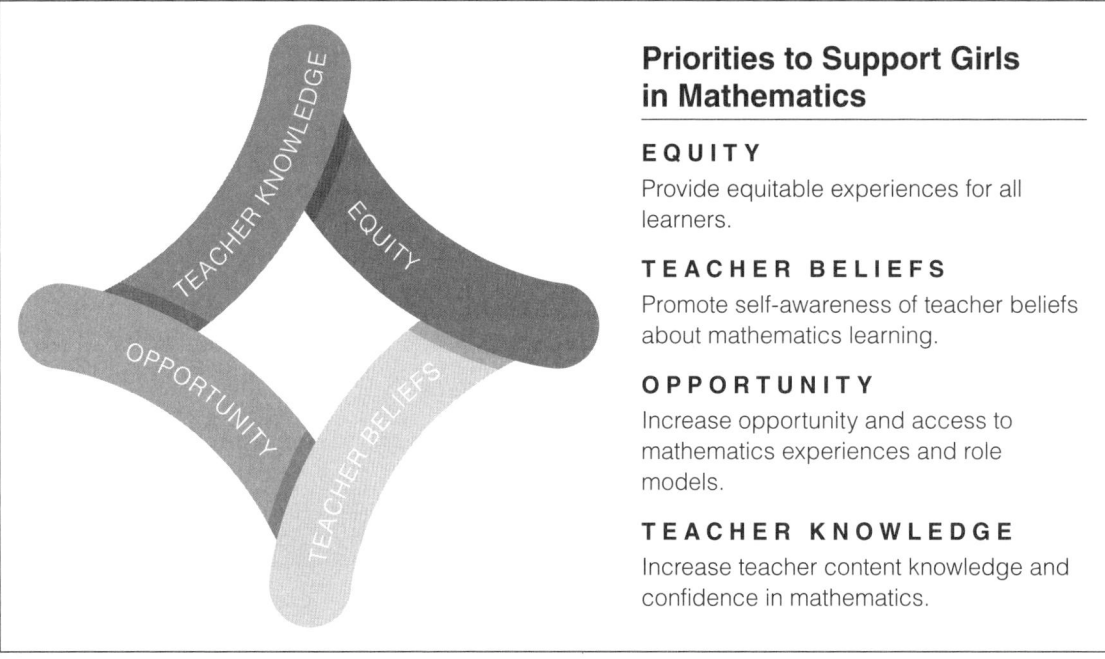

Figure 4.2: Priorities to support girls in mathematics.

The text that follows further clarifies these priorities.

Equity

Equity is often confused with the concept of equality. To clarify for our purposes, we will consider equity to mean *fairness*, whereas equality means *sameness*. Specifically, NCTM (2014) calls equity an assurance that all students have access to high-quality curriculum, instruction, and the supports that they need to be successful. We know that students enter the classroom with varied learning experiences, cultural backgrounds, and knowledge, which you must consider when creating or selecting mathematics tasks for students (Gutierrez, 2002; NCTM, 2000). Consider kindergarten students who enter school with limited oral language development and exposure. These young learners may need additional support or prompting in order to feel comfortable explaining their mathematical thinking and reasoning or to ask questions of their teacher or their classmates. It would be critical, however, to empower these learners through their mathematical learning interactions and class discussions, rather than to oversupport or exempt them with teacher thinking and language. It is critical for students to experience equity, or fairness, in the classroom, even though this may mean that they do not have identical experiences or supports.

Careful reflection and monitoring of equity are necessary, however, to ensure that fairness persists when differentiation of instruction is expected in our classrooms. Sometimes, even our well-meaning actions can be a roadblock for equity. Let's consider two such examples that will help us unpack this idea further.

- **Example 1:** If a student has a disability, such as Down syndrome, and is mainstreamed in the classroom, it will be necessary to consider particular instructional needs this student has. What is best for the student might be working with just one written mathematical task to avoid distraction or over-stimulation. Would this be best for all of the students in the class to keep everyone on the same pace or doing the same thing? Perhaps so. Perhaps not. Would it be fair for all of the students in the class? Perhaps so. Perhaps not. Therefore, this is an instructional model that a teacher will need to evaluate to determine what would be an equitable practice so that this student and all students are being well served in their mathematics learning experiences.

- **Example 2:** Consider that some students might be advanced mathematically. By consistently calling on these students during mathematics lessons, teachers are certainly reinforcing these students' mathematical understanding and growth. In addition, it shows that there is some learning taking place and some students are moving forward. But what is happening for the students who might not have readily available answers to problems or questions? What can teachers do to provide students with equitable opportunities to engage, even when they may not be ready to respond with correct answers?

As you think about the learners in your classroom who are girls, you do not necessarily need to plan for differentiated learning experiences by gender; however, you can use the same reflection questions that were offered in the previous two examples as a self-check to ensure that you are providing equitable experiences for all learners.

With careful consideration of each student's learning and development, in conjunction with high expectations for all learners, you can create individual pathways that offer equitable learning opportunities for mathematics success. You must reflect on the learning environment that you create and carefully examine your own preferences and tendencies for setting up mathematical learning experiences (whether they are explicit or implicit). This learning environment should be authentic to the needs of the students. For example, minimize examples and scenarios that rely on things you like personally such as shopping, sports, or crafting. In addition, do not falsely assume girls in your class like cooking, makeup, or stereotypically gender-based activities. Have discussions with your students to determine their interests and use examples that are real and relevant to their everyday lives and experiences. Because every student is different, you cannot provide a rigid, stringent learning environment and expect all students to have academic success. With this consideration, you would need to embrace equity to provide fair and just learning experiences for all students in the classroom.

According to the National Council of Teachers of Mathematics (NCTM, 2014), the following are productive beliefs in regard to equity.

- Students attain equity when they receive the differentiated supports (such as time, instruction, curricular materials, and programs) necessary to ensure that they are all mathematically successful.

- Equity applies to all settings.

- All students are capable of making sense of and persevering in solving challenging mathematics problems, and teachers should expect them to do so. Students, regardless of gender, ethnicity, and socioeconomic status, receive the support, confidence, and opportunities to reach high levels of mathematical success and interest.

Readers can use figure 4.3 to list ways their classrooms are equitable spaces where girls can learn mathematics.

 What are some strategies you use to promote equity in your classroom so that girls can learn mathematics at high levels?

Figure 4.3: Educator reflection on equity in the classroom.

As you seek to create equitable spaces, let's further explore the role of equity in girls learning mathematics through the lens of making an impact and using the TQE process as a guide for inclusiveness.

Making an Impact

As we consider equity in the classroom, think specifically about gender inclusiveness as it relates to your mathematics lessons. Consider a specific lesson that you recently taught. Did it include and equally represent all genders in your lesson? Consider the following four questions to assess the gender inclusiveness (or areas for potential improvement) for that lesson.

1. Does the lesson have language, tasks, examples, or other elements that are in any way demeaning or dismissive of girls and their interests? For example, does the lesson include roles that may be seen as gender based, such as cooking and cleaning? Or does the lesson focus on activities that are traditionally attractive to the boys in your class such as sports? Do you provide tasks with contexts that girls may not have been exposed to such as tying a tie or shaving facial hair? Will girls in your class have authentic engagement points?

2. Consider if you show preferences (either implicitly or explicitly) for certain approaches to mathematical problem solving that may give boys an advantage compared to girls. For example, do you ask for alternative approaches (using questions such as, Who used a different strategy that they might like to share?) consistently after responses offered by boys and girls? Similarly, do you follow up (using questions such as, How do you know? or statements like Show us how you were thinking?) consistently after students share both correct and incorrect strategies or solutions to generate productive discourse around mathematical thinking?

3. In the interest of engaging all students, does the lesson offer superficial gender elements as a claim that the lesson is pro-girl? For example, are you falsely positing narratives as it relates to girls and mathematics just to make the lesson work?

    - **Initial task:** Mark had some toys. His friend gave him 6 toys; now Mark has 13 toys. How many toys did he start with?
    - **Superficial revision:** Kyla had some toys. Her friend gave her 6 toys; now Kyla has 13 toys. How many toys did she start with?

4. Are you making a purposeful effort in designing the lesson to be inclusive (such as providing ample opportunity for students to ask and answer questions, addressing contexts with real-life focus, and so on) of all students in the classroom?

Using the TQE Process as a Guide for Inclusiveness of All Students

In order to impact girls' learning in mathematics, you must consider your instructional practices (Campbell et al., 2014) and aim to find ways to improve them. The TQE process (figure 4.4) can serve as a guide to improve the three essential components of every lesson: the task, the questions, and the evidence.

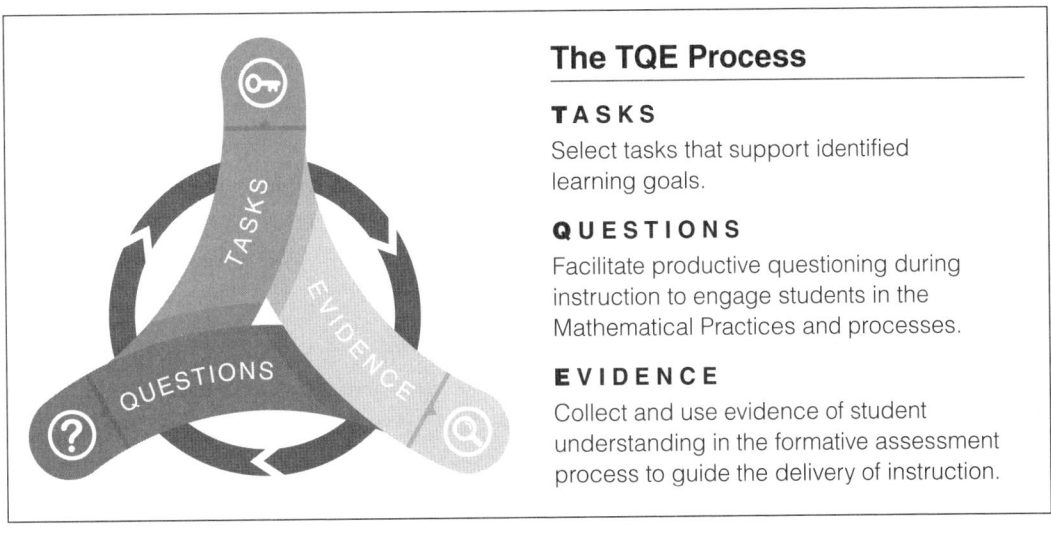

Source: Dixon, Nolan, & Adams, 2016, p. 4.

Figure 4.4: The TQE process.

We previously discussed tasks and questions during our reflections on the classroom videos in chapter 3 (page 33). Here, we will spend time on evidence and provide guidance for you to unpack an upcoming lesson. Retrieve your instructional materials and identify the upcoming standard you will teach. Using the standard and your instructional materials, review the possible tasks. As it relates to the task:

> Teachers with a deep understanding of the content they teach select tasks that provide students with the opportunity to engage in practices that support learning concepts before procedures; they know that for deep learning to take place, students need to understand the procedures they use. (Dixon, Nolan, Adams, Brooks, & Howse, 2016, p. 5)

As you review your instructional materials, notate the task or tasks that support meaningful student learning. Based on the notated tasks, ask yourself, "Will this task support the standard and help all students make connections in mathematics? Why or why not?" If you determine that it will not, consider how you will modify the task. Furthermore, you should ask three questions about the tasks you select.

1. "How are the tasks you are selecting encompassing and inclusive of the genders in your classroom?"
2. "Might a gender feel marginalized based on the task selected?"
3. "Will a gender be stereotyped based on the task selected?"

To prepare all students to benefit from the tasks you select, tasks should do three things.

1. Engage students in procedures with connections, doing mathematics, or both (Smith & Stein, 2011).

2. Build on the knowledge and experiences that students bring into the classroom (Carpenter, Fennema, Franke, Leui, & Empson, 2015).

3. Include the cultural background of students (Howse, 2013).

Now let's shift our focus to the questioning that will engage learners in discourse about their mathematical learning while assessing and advancing their thinking. Questioning is of critical importance as educators need to ensure that they ask all students questions in an equitable way. Questions should not be gender biased, whether purposefully or unconsciously. This involves both the content of questions and the patterns of questioning and facilitation in the classroom. For example, you might unknowingly offer probing questions about a strategy that one group of students in your classroom used (let's say that they happen to be boys), therefore implicitly devaluing the strategy that another group used (who may be girls). It is also important to ensure that all learners have an equal opportunity to ask questions and respond to questions asked of them. Questioning focuses on the "facilitation of targeted and productive questioning strategies that have a clear sense of how the content progresses within and across grades" (Dixon, Nolan, Adams, Brooks, & Howse, 2016, p. 6). As you anticipate how students may respond to the task or tasks you've selected, think about the questions that you imagine you might ask. Or, you might take a moment to think about the questions that you typically ask your students as you interact with them during mathematics instruction. Perhaps take a moment to write these questions down so you can refer back to them.

Consider the purpose of each question that you ask your students. How do your questions ascertain student understanding? Do your questions advance students' mathematical thinking? How do your questions challenge your students as they do mathematics, but simultaneously offer support and confidence that they can achieve? Through questioning, work to ensure all students receive a variety of questions that allow them to engage in mathematics in meaningful and challenging ways. For more examples of questions that you might build into your lesson, revisit chapter 3 (page 33) to read more about assessing and advancing questions.

Finally, let's think about the student evidence of mathematical learning that you gained from a recent lesson. The use of evidence gained from the formative assessment process during mathematics instruction can help you "know where to linger in developing students' coherent understanding of mathematics" (Dixon, Nolan, Adams, Brooks, & Howse, 2016, p. 6). As you consider evidence of mathematical learning in your own lesson, think back to the learning goal for your selected task or tasks. Based on the learning goal, think about what you were specifically seeking for students to accomplish through the task. Consider how you will assess individual student's mathematical thinking—especially girls'—and progress toward the learning goal. Determine whether or not you have a clear and concise method to communicate with students about their progress based on their individual evidence. There is not one right answer about what evidence to collect and how to use evidence as a springboard for future instruction; however, engaging in intentional reflection about student evidence is a powerful way to improve instruction. When you purposefully ensure that you are equitable about collecting student evidence, you provide opportunities for girls (as well as boys) to demonstrate what and how they are learning.

Use figure 4.5 to reflect on ways you can communicate clearly to students.

 What can you plan to do to make sure you communicate clearly to girls about their progress in mathematics?

Figure 4.5: Educator reflection on ways to communicate student progress to girls.

Teacher Beliefs

Students bring into the classroom a wealth of knowledge and experiences (Carpenter et al., 2015). As educators, it is our responsibility to build on this. Think about what experiences students bring into your classroom setting. What are some of their strengths? What are some of their weaknesses?

According to Campbell et al. (2014), teachers' beliefs and awareness of students' prior mathematics experiences directly impact their instructional practices. In turn, teachers' access to professional development, their own professional background, and their former teaching experiences (Campbell et al., 2014) also influence their beliefs. Joseph R. Cimpian, Sarah T. Lubienski, Jennifer D. Timmer, Martha B. Makowski, and Emily K. Miller (2016) suggest that "stereotypes related to gender and mathematics persist" (p. 3). In other words, there are still people who perceive boys at being better in mathematics than girls just because the boys are boys, and in fact, girls, boys, teachers, parents, and others often hold this perception. Cimpian and colleagues (2016) find that, in mathematics, "teachers give lower ratings to girls when boys and girls perform and behave similarly . . . [and] . . . perceive girls as working harder than similarly achieving boys in order to rate them [girls] as similarly proficient in math" (p. 16). With this understanding, it is imperative to call attention to the direct and indirect impacts of teacher beliefs on instructional practice and the need for increased self-awareness of teachers' own beliefs. Think more about these research findings presented in figure 4.6.

 What is your reaction to the findings of Cimpian and colleagues (2016) that, in mathematics, "teachers give lower ratings to girls when boys and girls perform and behave similarly . . . [and] . . . perceive girls as working harder than similarly achieving boys in order to rate them [girls] as similarly proficient in math" (p. 16)?

Figure 4.6: Reaction to research findings on teacher perceptions of boys' and girls' classroom performance.

In this section, we will offer two tools that you can use to increase your own awareness about mathematics teaching and learning and your own perceptions about providing mathematics learning experiences for girls in your classroom (which we believe may be evolving from the perceptions you held prior to starting this book). Use these tools to take an assessment of your beliefs and reflect upon the results.

For each item in figure 4.7 (page 74), consider the statements adapted from NCTM's (2014) *Principles to Action*, and select the statement that you agree with in each line. Read both statements thoroughly, and honestly respond with the statement that resonates with your beliefs. Do not focus on selecting the

Line 1	Mathematics learning should focus on practicing procedures and memorizing basic number combinations.	Mathematics learning should focus on developing understanding of concepts and procedures through problem solving, reasoning, and discourse.
Line 2	All girls need to have a range of strategies and approaches from which to choose in solving problems, including, but not limited to, general methods, standard algorithms, and procedures.	Girls need only to learn and use the same standard computational algorithms and the same prescribed methods to solve algebraic problems.
Line 3	Girls can learn mathematics through exploring and solving contextual and mathematical problems.	Girls can learn to apply mathematics only after they have mastered the basic skills.
Line 4	The role of the teacher is to tell girls exactly what definitions, formulas, and rules they should know and demonstrate how to use this information to solve mathematics problems.	The role of the teacher is to engage girls in tasks that promote reasoning and problem solving and facilitate discourse that moves students toward shared understanding of mathematics.
Line 5	The role of the student is to be actively involved in making sense of mathematics tasks by using varied strategies and representations, justifying solutions, making connections to prior knowledge or familiar contexts and experiences, and considering the reasoning of others.	The role of the student is to memorize information that the teacher presents and then use that information to solve routine problems on homework, quizzes, and tests.
Line 6	An effective teacher provides girls with appropriate challenges, encourages perseverance in solving problems, and supports productive struggle in learning mathematics.	An effective teacher makes the mathematics easy for girls by guiding them step by step through problem solving to ensure that they are not frustrated or confused.

Figure 4.7: Belief statements about mathematics teaching and learning.

*Visit **go.SolutionTree.com/mathematics** for a free reproducible version of this figure.*

"right" statement but rather the selection of the statement that is personal to you. Upon completion, reflect on the statements and consider why you selected each. How do your selections reflect your beliefs about teaching and learning mathematics? Do you think your selections match your teaching practices in your own classroom? As you communicate with fellow teachers, school leaders, district leaders, and other stakeholders, it will be important for you to have already thought about the way you will support girls in mathematics. Consider other helpful positions you might take. These items will help you plan your responses.

Now, take the self-reflection quiz in figure 4.8 to revisit your own perceptions about teaching mathematics for all students, with an emphasis on your perceptions about girls in your classroom and the experiences that you provide for them.

Select a choice for each of the thirteen statements.

1. During mathematics instruction, I provide the same opportunities for boys and girls.
   a. Strongly agree
   b. Agree
   c. Disagree
   d. Strongly disagree

2. Mathematics is a male-dominated field.
   a. Strongly agree
   b. Agree
   c. Disagree
   d. Strongly disagree

3. I believe my girls can be successful in mathematics.
   a. Strongly agree
   b. Agree
   c. Disagree
   d. Strongly disagree

4. I believe it is harder for girls to learn mathematics.
   a. Strongly agree
   b. Agree
   c. Disagree
   d. Strongly disagree

5. The examples I use in class engage my girl students.
   a. Strongly agree
   b. Agree
   c. Disagree
   d. Strongly disagree

6. I praise boys more frequently than I praise girls.
   a. Strongly agree
   b. Agree
   c. Disagree
   d. Strongly disagree

7. The girls in my classroom like mathematics.
   a. Strongly agree
   b. Agree
   c. Disagree
   d. Strongly disagree

8. I am adequately trained to provide an inclusive mathematics environment for girls.
   a. Strongly agree
   b. Agree
   c. Disagree
   d. Strongly disagree

9. I have a good rapport with girls in my classroom.
   a. Strongly agree
   b. Agree
   c. Disagree
   d. Strongly disagree

10. I provide tasks that highlight women in STEM roles.
    a. Strongly agree
    b. Agree
    c. Disagree
    d. Strongly disagree

11. I work to ensure that girls see themselves in future STEM roles.
    a. Strongly agree
    b. Agree
    c. Disagree
    d. Strongly disagree

12. I believe in myself as a mathematics teacher.
    a. Strongly agree
    b. Agree
    c. Disagree
    d. Strongly disagree

13. I struggle when teaching mathematics.
    a. Strongly agree
    b. Agree
    c. Disagree
    d. Strongly disagree

Figure 4.8: Questionnaire for revisiting your perceptions about teaching mathematics to girls.

Visit go.SolutionTree.com/mathematics for a free reproducible version of this figure.

As you reflect on each item, do you notice a difference in your responses compared to similar questions we presented you with previously (page 20)? Did you have strong responses (either positive or negative) as you were reflecting on each, or did you find yourself to be more neutral? Lastly, as you consider questions eight through eleven, do you think the girls in your classroom have opportunities and access to learn about STEM-related fields, to learn from role models (including you) who demonstrate mathematical power, or both? In the next section, we will explore each of these ideas further.

Opportunity

Hill et al. (2010) posit three main barriers to girls in pursuing STEM-related careers: (1) the notion that boys and men are intellectually superior in regard to STEM fields, (2) girls' lack of interest in STEM, and (3) STEM work-life balance bias. First, Else-Quest et al. (2010) find that boys and girls differ very little in mathematics achievement, debunking the notion of gender superiority. They state that "females perform at the same level as male classmates when they are encouraged to succeed, are given the necessary educational tools, and have visible female role models exceling in mathematics" (Else-Quest et al., 2010, p. 125). Second, we have already presented research that refutes the suggestion that girls lack interest in STEM-related fields. In fact, we present evidence that suggests the majority of teenage girls have interest in STEM-related careers, despite awareness of barriers and biases based on their gender alone (Girl Scout Research Institute, 2012). Steven J. Spencer, Claude M. Steele, and Diane M. Quinn (1999) find that negative stereotypes hinder girls when completing mathematics tasks, which can impact girls' desire and willingness to engage in such tasks. This leaves the third barrier: STEM work-life balance bias. Data from the Girl Scout Research Institute (2012) reveal that, of those girls who expressed interest in STEM, only 13 percent say it is their first choice as a career. Interestingly, 30 percent of those girls are interested in being stay-at-home mothers, suggesting that even at a young age, many females are already considering work-life balance when pursing STEM pathways. How can we increase opportunities and access for our young girls to not only develop and maintain an interest in mathematics (and thereby STEM-related fields) but also see and learn from examples and role models who have experienced success as underrepresented women? Consider the quote in figure 4.9 about representation.

 How would you explain the quote "It's hard to be what you can't see" (Edelman, 2015) as it pertains to girls in mathematics?

Figure 4.9: Reaction to quote on the representation of girls in mathematics.

We have learned that increasing the quality of girls' mathematical learning experiences and the role models in their support networks can, in itself, have a significant impact on girls' mathematical achievement and career paths. The following are four concrete examples to increase opportunities and access for girls, while also promoting positive perceptions about the success of girls in mathematics.

1. Sponsor mathematics honor societies or clubs for girls or those that are strongly inclusive of girls and hold events that are open to the public. You can consider organizations such as Girls Who Code (https://girlswhocode.com), Robotics clubs, Odyssey of the Mind (www.odysseyofthemind.com), Math Olympiads for Elementary and Middle Schools (www.moems.org), to name a few. (Visit **go.SolutionTree.com/mathematics** to access live links to the websites mentioned in this book.)

2. Request local agencies that serve children and that also have a national presence, such as Boys & Girls Clubs of America (www.bgca.org), to sponsor programs that promote girls in mathematics.

3. Invite guest speakers (such as accountants, meteorologists, and so on) from the community to mentor girls in mathematics. This is to benefit the girls by increasing their access to role models, but it also gives community members opportunities to engage with girls studying mathematics.

4. Collaborate with colleges and universities in writing grants to support girls studying mathematics. Grants such as those NCTM (www.nctm.org/Grants) and the Mathematics Association of America (www.maa.org/programs/maa-grants) offer are a place for you to start.

Thinking about how to impact society regarding perceptions about girls studying mathematics might seem like a big hurdle. However, we can engage in opportunities to inform stakeholders in society, such as local community leaders and civic and educational organizations about girls in mathematics.

Public school teachers in the 21st century climate are increasingly involved in the politics of education. Thus, teachers need more awareness of how policies develop and are implemented at the school, district, state (or province), and national levels. One must understand the main educational governing force is the federal government. The federal government establishes policy, collects data, and ensures equal access to education.

Therefore, there is further need to delineate governance and policy. Paul Manna and Patrick McGuinn (2013) define *governance* as the "process by which formal institutions and actors wield power and make decisions that influence the conditions under which people live in a society" (p. 9). They define *policy* as "the array of initiatives, programs, laws, regulations, and rules that the governance system chooses to produce" (Manna & McGuinn, 2013, p. 9). Understanding these two terms allows one to make better sense of the differences. An example of a policy that specifically affects girls is tracking. Tracking in early grades comes in the form of ability grouping, and in the later grades involves separating students into different mathematics courses based on their prior performance. According to the Brown Center on Education Policy (2013), approximately 75 percent of U.S. students are tracked. These tracking policies can impact girls as if they are in a "lower" track because it may impact their ability to take advanced courses during their secondary school curriculum, thereby further exacerbating the STEM gender gap in postsecondary education and careers. After considering this issue for its students, the San Francisco Unified School District implemented a policy to "de-track" students and provide all students the opportunity to gain a solid mathematics foundation, which the district believes will better prepare them for future challenging mathematics courses (Sawchuk, 2018). De-tracking policies can be instituted at the federal, state, district, and school levels.

As teachers, you are on the front line of the educational process, and you directly impact students and student achievement. Perhaps there are some education policies you would like to change that would provide better opportunities for girls to excel in mathematics. What are some of these policies?

The following are six steps (Community Tool Box, 2018) to organize how to address policies that impact girls studying mathematics in your school.

1. **Learn about a current policy:** To accomplish this step, one needs to review the current policies in place. For example, in the tracking policy example we previously shared, this step may involve learning more about the process of tracking in general, including some background information.

    The tracking of students typically begins with screeners (that is, achievement assessments). These screeners are designed to determine students' ability in mathematics, which in turn determines placement. In some instances, assessment measures are not used and placement is based upon recommendations from counselors, teachers, parents, or administrators. Tracks are typically based upon three levels: (1) below level students, (2) on level students, or (3) above level students. This allows the homogeneous grouping of students, which lends itself to instruction customized to meet student needs. In the elementary level, tracking is in the form of ability grouping, and in the secondary level, tracking is in the form of course assignment.

2. **Identify a policy to change:** This step occurs after reviewing the current policies. In identifying a policy to change, one should also weigh the pros and cons that would be outcomes of the change. In the tracking policy example, this step would involve clearly identifying the change that one desires: the de-tracking of students' curricular progression in order to provide all students an equal opportunity to access high-quality mathematics instruction, tasks, and courses.

3. **Solicit support:** Seek support from internal colleagues, administrators, community members, and key stakeholders. In the case of the tracking example, support would need to be solicited from many stakeholders.

    - *Parents*—Their children would be directly impacted.
    - *Teachers*—They will need to be well versed in differentiated instruction and inclusive pedagogical strategies to meet the needs of learners at a wide range of levels.
    - *Administrators, school counselors, and psychologists*—They will need to understand class loads, psychological impacts for learners, and school achievement data.

4. **Research alternatives:** Find out about alternatives to the policy in collaboration with the stakeholders. One must clearly understand and be able to articulate the issue, including possible alternatives and their validity. It can be especially helpful for individuals seeking policy change to work with others, potentially including colleagues, teams or groups, unions

or other organizations, and even elected officials, for support. In the policy change tracking example, alternatives generated may include—

- Leaving the current tracking policy in place
- Having a pilot group to test the de-tracking policy
- Providing both tracking and nontracking options
- Improving current remediation policies (which has implications for the relevance of tracking policies)

5. **Develop an action strategy including solid rationale:** The previous steps build toward the development of the action strategy and rationale. In the case of the tracking policy example, one would need to collect strong research about the pros and cons of de-tracking. He or she would need to synthesize and share this research with stakeholders in both the synthesized (such as a white paper) and full report form. Focus groups would need to discuss the current policy, the proposed policy, and alternatives, while assessing short- and long-term consequences. Stakeholders should maintain a student-centered focus during this process. In addition, they would need to enact parameters to ensure equitable information and representation are in place as they make decisions.

6. **Present the proposed policy change:** The intended audience of this presentation depends on the level at which the policy decision is being made (for example, a school board meeting, letter to or meeting with a congressman, or even a congressional hearing). After robust discussion and gathering the sufficient data, a final plan of action would be submitted detailing the policy change. Considering the tracking policy example, the proposed policy change was decided upon by the San Francisco Unified School District as the governing body to change the policy for schools.

The key with change is starting. As educators, we cannot continue to assume change is going to occur without our contribution to the process. As everyday practitioners, we have a strong sense of the nuances of teaching. Thus, it is up to us to lead the charge in rectifying these issues. By using the steps to implement change, you are challenged to begin implementing change in your respective environment.

Teacher Knowledge

Scholars have debated for years the mathematical knowledge needed for teaching (Ball, Hill, & Bass, 2005; Skemp, 1978); however, there is currently a strong call in both research and practice to increase the content knowledge of teachers who are teaching mathematics (Dixon, Nolan, Adams, Brooks, & Howse, 2016). For example, Deborah Loewenberg Ball, Heather C. Hill, and Hyman Bass (2005) state, "How well teachers know mathematics is central to their capacity to use instructional materials wisely, to assess students' progress, and to make sound judgments about presentation, emphasis, and sequencing" (p. 14). When teachers lack a deep understanding of the content they are teaching, their ability to teach meaningful lessons is inhibited, regardless of the quality of the curriculum or instructional materials they are using (Campbell et al., 2014; NCTM, 2014). Early research echoes these points. Richard R. Skemp (1978) says:

The depth of mathematical knowledge needed in order to effectively teach is further determined by the level of understanding one is seeking to impart in students, whether it be relational understanding (focusing on the what and why) or instrumental understanding (applying rules without reason). (p. 9)

In order to assist teachers in increasing their content knowledge, seek out high-quality resources. Resources that have been vetted and that are useful to help teachers can impart a depth of understanding of the mathematics they are teaching and develop an understanding of the students they are instructing. Collaboratively work to identify purposeful and meaningful resources. Develop a repository of this information to use with future teachers. In addition, identify customized professional learning opportunities to provide teachers with additional exposure to experts and resources.

As you consider the two types of understanding, ponder which is more likely to focus on girls. Does your education environment challenge girls to think critically, or does it probe them to focus on processes, procedures, and obtaining correct answers? As you might assume, our goal in providing high-quality mathematics for all learners is to build relational understanding, thus warranting a deep content knowledge base for those who teach mathematics. In order to ensure that all students receive a high-quality mathematics learning experience, Heather C. Hill and Sarah Theule Lubienski (2007) posit equitably distributing knowledgeable teachers, increasing the efficiency of educator preparatory programs, and genuinely focusing on teacher quality.

Let's just pull away for a moment and visit *Making Sense of Mathematics for Teaching Grades K–2* (Dixon, Nolan, Adams, Brooks, & Howse, 2016), which includes the core principle of giving attention to "a progression of learning for a big idea" (p. 8). We strongly believe that when educators explore and understand such progressions, they are able to make sense of mathematics for teaching that empowers students to learn conceptually as well as procedurally.

For example, here is an outline of the progression for introducing basic operations:

- Make sense of contexts that support addition and subtraction where the result of adding or subtracting is unknow.
- Make sense of contexts that support addition and subtraction with unknowns in all positions.
- Solve two-step word problems involving addition and subtraction with unknowns in all positions.
- Make sense of addition situations involving adding equal groups to prepare for multiplication.
- Make sense of contexts that support multiplication and division. (Dixon, Nolan, Adams, Brooks, & Howse, 2016, p. 36)

Teachers who have the opportunity to deeply explore such progressions learn about mathematical examples and non-examples, anticipated errors students might make, and mathematical connections that all contribute to meaningful and effective instruction. This is one way teachers can build their knowledge of mathematics across the curriculum and for any grade level or grade band.

Use figure 4.10 to list your own recommendations for high-quality learning.

> Review Hill and Lubienski's (2007) recommendations for high-quality learning. What are your reactions to this list? Is this a comprehensive list, or are there other recommendations that you would include?

Figure 4.10: Recommendations for high-quality learning.

An additional outcome of increased teacher content knowledge is an increase in teachers' own self-confidence as they make sense of the mathematics they are teaching. As we discussed in chapter 1 (page 7), research shows a link between women teachers' anxiety about mathematics and its influence on girl students.

We also know that the early perceptions that girls develop about studying mathematics matter. Cimpian et al. (2016) suggest that "math achievement in elementary school appears to influence girls' emerging views of mathematics and their mathematical abilities" (p. 2). These emerging views impact decisions girls make about studying mathematics as they progress in school. Thus, the opportunity to learn from a knowledgeable, confident teacher of mathematics in the early years of schooling cannot be understated.

What is strongly needed is for you to have solid knowledge of mathematics for teaching. Confidence comes along with that. We refer you to the appropriate grade band of *Making Sense of Mathematics for Teaching* to support a deep dive into mathematics content (Dixon, Nolan, Adams, Brooks, & Howse, 2016; Dixon, Nolan, Adams, Tobias, & Barmoha, 2016; Nolan, Dixon, Roy, & Andreasen, 2016; Nolan, Dixon, Safi, & Haciomeroglu, 2016).

Use figure 4.11 to brainstorm ways to promote teacher self-efficacy.

> How can you promote teacher self-efficacy in your school district?

Figure 4.11: Ways to promote teacher self-efficacy in the school district.

Conclusion

This chapter focuses on the role of equity, teacher beliefs, opportunity, and teacher knowledge as priorities to strengthen the experiences of girls in mathematics classrooms. As you considered each of these priorities, where did you see opportunities to advocate for girls' learning, both in your classroom and beyond? In what ways might you have work to do to strengthen your content knowledge and confidence as a mathematics teacher? How might you get involved in learning about and advocating for policies that increase girls' opportunities in mathematics during and beyond the school day? We hope that continuing to think about these four priorities will help you provide the experiences, tools, and support networks to empower all learners, but especially the young girls that you teach.

Figure 4.12 (page 82) offers actions to implement priorities that teach mathematics to girls.

> The following three actions can help you implement priorities to teach girls mathematics.
>
> 1. Consider using or modifying the TQE process to develop your next gender inclusive lesson. As you develop this lesson, consider your student population and the interests of all your students. The key is to formulate a lesson that encompasses student interests. Thus, it is imperative that you genuinely talk to your students and get to know your students. Identify their likes and dislikes. Learn what fascinates them. This keen understanding of your students will assist in the development of rich and rewarding lessons.
> 2. Brainstorm with your students how your lessons can expand beyond the traditional mathematics instructional time block. While students making sense of mathematics inside the classroom is a goal, students being able to apply that knowledge outside the classroom to real-world situations is even better. Thus, consider how the mathematics lessons can benefit the school, community, or both. Consider the incorporation or use of project-based learning to build on singular classroom lessons to have unit-based inquiry lessons.
> 3. Take the time to research a piece of recently passed legislation related to education in your state (or province). Consider your personal and professional reactions to the legislation. Find time to share what you have learned with colleagues, and consider how you might reach out to policy makers in your state to share your reactions and recommendations.

Figure 4.12: Actions for implementing priorities in teaching girls mathematics.

Reflections

Answer the following five questions independently or in your book study to further your understanding and goals related to teaching girls mathematics.

1. As you reflect on the TQE process, consider some tasks that you have recently selected for your students. Did the tasks you assigned pique your students' interests? How do you know?
2. Consider your dual role as teacher and role model in the mathematics classroom. What are five things you know about your students that will enhance your relationship with your students, specifically girls, in your classroom? How can you help your students see you as a part of their support network toward exploring advanced paths in mathematics?
3. What are some efforts being implemented in your educational environment to ensure all teachers have mathematical content knowledge and are improving on this knowledge? What are your most recent efforts to increase your own mathematics content knowledge and confidence?
4. How can you express your confidence in knowing and teaching mathematics to serve as a positive role model for students?
5. What are three things you will do to improve your mathematics instruction?

Further Reading

Dennehy, T. C., & Dasgupta, N. (2017). Female peer mentors early in college increase women's positive academic experiences and retention in engineering. *Proceedings of the National Academy of Sciences, 114*(23), 5964–5969.

Kiwanuka, H. N., Van Damme, J., Van Den Noortgate, W., Anumendem, D. N., Vanlaar, G., Reynolds, C. et al. (2017). How do student and classroom characteristics affect attitude toward mathematics? A multivariate multilevel analysis. *School Effectiveness and School Improvement, 28*(1), 1–21.

Miller, A. D., Ramirez, E. M., & Murdock, T. B. (2017). The influence of teachers' self-efficacy on perceptions: Perceived teacher competence and respect and student effort and achievement. *Teaching and Teacher Education, 64*, 260–269.

Nurlu, Ö. (2015). Investigation of teachers' mathematics teaching self-efficacy. *International Electronic Journal of Elementary Education, 8*(1), 21–40.

Perez-Felkner, L., Nix, S., & Thomas, K. (2017). Gendered pathways: How mathematics ability beliefs shape secondary and postsecondary course and degree field choices. *Frontiers in Psychology, 8*, 386.

Salikutluk, Z., & Heyne, S. (2017). Do gender roles and norms affect performance in maths? The impact of adolescents' and their peers' gender conceptions on maths grades. *European Sociological Review, 33*(3), 368–381.

Steinke, J. (2017). Adolescent girls' STEM identity formation and media images of STEM professionals: Considering the influence of contextual cues. *Frontiers in Psychology, 8*, 716.

Stoehr, K. J. (2017). Mathematics anxiety: One size does not fit all. *Journal of Teacher Education, 68*(1), 69–84.

EPILOGUE
Encouragement for Girls in Mathematics

Girls are interested in STEM! Talk to them like they are.

—Girl Scout Research Institute

We realize that on a day-to-day basis, it can be quite challenging to motivate students to engage in mathematics in authentic ways. A question we receive quite frequently is, "What can I say to motivate girls or [insert any student group here] to learn mathematics?" This question is usually followed up with, "I run out of things to say (or do)!" So, we end *Making Sense of Mathematics for Teaching Girls in Grades K–5* with resources to encourage girls in mathematics, including some examples of things you can say and do as you strive to motivate girls to stay connected to and engaged with mathematics. Feel free to tweak and adapt what we share to make it a good fit for the girls (and any student) you teach.

Resources to Encourage Girls in Mathematics

When asked to talk about the life of an imaginary woman in a STEM career, a group of girls says (Girl Scout Research Institute, 2012):

> We think a woman in STEM works harder because she wants to get everything; that is her goal. She gets frustrated at work when the guys don't support her ideas, but it makes her work harder and she becomes stronger.
>
> You can do whatever you want if you work hard and persist—but you have to have the skills. You have to prove them wrong. Don't give up. If you keep saying yes to everything, it's not good. It's good to have a certain amount of resistance; good for someone to tell you "no," because that will happen in your life. . . .
>
> Barriers will make you stronger and persist. [Girls] persist because [they have] knowledge and dedication. Tired of hearing "no"—you will get told "no" a lot in your life. Just keep going and don't give up.

How would you respond to the preceding comments from the Girls Scout Research Institute? As the group states, "You will get told 'no' a lot in your life. Just keep going and don't give up." Use figure E.1 to reflect on a time when you were told no and your response to that.

Think of a time when someone told you "no." How did you respond? Were you discouraged? Did you persist? How do you think this and other experiences shaped your goals now?

Figure E.1: Reflection on learning from a "no."

The following are four resources you can use to encourage girls to persist in mathematics, even after hearing "no."

1. **Open letter to your girl students:** Consider sending an open letter to the girls in your classroom. See figure E.2 for an example. You might read this letter to students if needed and also send a copy home with each student. You may also print copies and have each student paste the letter on the front of her mathematics journal as a reminder of your advocacy for your students. Feel free to modify the letter to fit your class need so that it is developmentally appropriate for the grade or grades you are teaching.

> Dear Student,
>
> I am so excited that you are in my class and that I get to teach you mathematics! Maybe you've heard people say these things regarding mathematics.
>
> - "Mathematics is boring."
> - "Mathematics is hard."
> - "I hate mathematics."
> - "I can't do mathematics."
>
> Well here is what I have to say about those statements: "No. No. No. No."
>
> Here is what we are going to do: we are going to learn mathematics together and work on tasks that are meaningful. We are going to work on mathematics that will not only challenge you but also help you learn. I will show you what I love about mathematics, and I will tell you why I believe that you can do mathematics. Know and understand you will make mistakes. At times it will seem hard, but if you work hard and persevere, you will find success.
>
> When we are learning mathematics, there are just a few things I want you to do for yourself: Do your best. Try your best. Ask your best questions. Give your best answers. Share your best ideas. You may think sometimes that mathematics is hard—well, that's when we will have the most fun learning! Let's do some mathematics!
>
> Sincerely,
> Teacher

Figure E.2: Sample teacher letter of encouragement to girls.

Visit go.SolutionTree.com/mathematics for a free reproducible version of this figure.

2. **Open letter to you:** We have a letter for you too! Use the sample letter (see figure E.3) to inspire yourself on your journey to providing an inclusive environment for girls in mathematics. You may also print a copy and post it near your desk for continuous inspiration.

3. **Affirmations for your students:** Figure E.4 (pages 87–88) offers a list of one hundred affirmations that you can use in conversations with individual students, small groups, or the whole class. You can also consider posting these affirmations on students' desks and in other areas in the classroom. Girls benefit from authentic encouragement to excel; in fact, all students benefit from this! You might also consider allowing each student to pick an affirmation that is particularly meaningful and encouraging to her. The one thing we cannot do is give you the passion you need to make these statements more effective. We can give you the statements, but conveying the passion through the words is up to you.

Dear Teacher,

We are so excited that you are on this journey with us to provide an inclusive environment for girls in mathematics. While each of us has had different experiences teaching girls mathematics, we all agree they deserve an equitable learning experience—an experience that is inclusive of their interests and potential.

Take a moment and picture a current or former girl student. Think about just how much you care about this student. Now imagine she is under your purview; wouldn't you want this student to have the best mathematics experience possible?

Repeat the following affirmation daily to provide yourself with encouragement: "All my students have tremendous potential. It is up to me to get to know my students and unlock this potential. I will diligently work with girls to ensure their voices are heard in my classroom. I will purposefully seek out their interests and push for their potential. I will actively work to ensure girls are engaged in mathematics. I will believe in myself as a mathematics educator and role model. My students will feed off of my passion for the subject. Most important, I will create an environment in which girls can demonstrate their knowledge. I will give them opportunities for success in mathematics."

Sincerely,
The Authors

Figure E.3: Sample teacher affirmation letter.

*Visit **go.SolutionTree.com/mathematics** for a free reproducible version of this figure.*

1. Mathematics is for you (for everyone).
2. You (everyone) can learn mathematics.
3. You (everyone) can do mathematics.
4. As long as you try to do mathematics, you're doing mathematics.
5. We use mathematics every day.
6. I love mathematics. (State one to two reasons why.)
7. If something in mathematics doesn't make sense to you, it's okay to ask questions.
8. Mistakes in mathematics are normal. I make them too.
9. Getting wrong answers in mathematics is normal. You don't have to feel bad about that.
10. Just roll with the punches when you're working in mathematics. It's going to be all right.
11. No task is too hard to try.
12. Trust what you're thinking and give it a try.
13. Trust and share your ideas.
14. Put your thoughts out there.
15. Believe in yourself.
16. Believe in your ability to solve this task.
17. There can be many paths to success in mathematics.
18. Think. Try. Think again. Try again.
19. Tired of working on a task. Take a break, then get back to it.
20. Next time you try, you'll get it done (or right).
21. Never give up on yourself.
22. What matters most is that you think about it.
23. Your effort makes all the difference.
24. Make a plan. Stick to it until it doesn't work.
25. You can guess as long as you check.
26. Women mathematicians were once girls too.
27. You may be wrong; we are all wrong sometimes, but that's how we learn.
28. Strive to learn something new in mathematics class every day.
29. Your gift to the team (or group) is how you contribute to the task.
30. Lead with your mind.
31. Do whatever it takes (for example, use a tool, try another method) to solve a task.
32. Want it for yourself.

continued →

33. I believe in you; I need you to believe in you.
34. Focus.
35. What's your goal for this task? You can make it there.
36. As long as you stay in the race, you have a chance to win.
37. Never sell yourself short.
38. If you see someone struggling in your group, help, but don't give the answer away.
39. Slow does not mean low.
40. Be your own competition.
41. You have what it takes.
42. Be your own biggest fan.
43. Never doubt yourself.
44. You have what it takes to be great at this.
45. Be courageous.
46. You bring value to the table.
47. Not trying is the biggest barrier to success.
48. Always try one more time.
49. Nothing can stop you from learning mathematics.
50. Make yourself proud.
51. Don't let the tough tasks get you down.
52. Do not tell them. Show them how great you are!
53. Do as much as you can.
54. When you come to mathematics class, show up and shine.
55. Mathematics can't beat you.
56. Struggling is part of doing mathematics. Don't fear the struggle.
57. If anyone can do mathematics, you can.
58. Little by little, step by step, you can do mathematics.
59. See yourself doing mathematics.
60. Mathematics is your friend.
61. I will be the next _____!
62. I can use mathematics to change the world.
63. I am a problem solver.
64. Mathematics is not just for the boys.
65. I will be successful.
66. Nothing happens if I don't try
67. I can do anything in mathematics.
68. I will not let anything stop me from achieving my goal.
69. Showing my work shows my thinking.
70. Appreciate the beauty of mathematics.
71. We can use mathematics for work and play.
72. Let no one stop you on your path to greatness.
73. Your effort makes a difference.
74. There isn't anything that can stop you.
75. Keep moving forward.
76. Mathematics can change your life.
77. Mathematics opens new doors of opportunity.
78. You are surrounded by mathematics.
79. Just know every day you learn more and more in mathematics.
80. Today is a good day to learn mathematics.
81. Celebrate your mathematics mind.
82. See the beauty in mathematics.
83. Let the beauty of mathematics inspire you.
84. Mathematics can take you places.
85. Relax, you can do this.
86. Be limitless, like mathematics.
87. If you look at mathematics positively, you have a better time learning it.
88. Believe in the power of mathematics.
89. It's okay to start over.
90. Mathematics has stumped some famous mathematicians too.
91. Mathematics opens a world of possibilities.
92. Having a positive attitude in mathematics helps a lot.
93. Your work is amazing (or brilliant or spectacular).
94. Science needs mathematics.
95. Continue working hard.
96. Everybody can become a problem solver.
97. Getting to the solution is not the end of learning mathematics.
98. Let mathematics start your adventure.
99. Go for it!
100. The joy of mathematics can be yours

Figure E.4: One hundred mathematics affirmations for students.

*Visit **go.SolutionTree.com/mathematics** for a free reproducible version of this figure.*

4. **Pick-a-word student activity:** Present your students with a list of words such as those in figure E.5 that might convey how they feel about mathematics and how they feel as learners of mathematics. Instruct students to pick a word and then use the word to present a description. This will provide an opportunity to reinforce positive perceptions and combat negative perceptions. You should add to this list yourself!

Directions: Students will select the word from the following word bank that best describes how they feel about mathematics. Then, they'll share why they selected that word. The mathematics teacher or coach can use students' responses to kick-start a conversation about mathematics and to combat any negative perceptions.

Afraid	Usual	Confident
Angry	Fast	Doubtful
Serious	Sad	Like
Nervous	Low	Calm
Normal	Anxious	Courageous
Hate	Weak	Different
Happy	Love	Same
High	Slow	Proud
Hopeful	Alone	Hard
Strong	Handy	Smart

Figure E.5: Pick-a-word student activity options.

Visit **go.SolutionTree.com/mathematics** *for a free reproducible version of this figure.*

Conclusion

As we conclude this book, we leave you with a belief. A belief that this book has helped you increase your knowledge of teaching girls mathematics. A belief that your perceptions of teaching girls mathematics have been positively impacted. A belief that you understand the possibilities of girls learning mathematics and will inspire others to become better teachers of girls in mathematics. A belief that you have confidence in yourself to prioritize teaching girls mathematics in order to bring about change to girls' mathematics experiences. This change starts with a singular action that can have a domino effect to facilitate larger change. Trust and believe in yourself. We challenge you to consider one thing you are going to do to bring about change to the way girls experience mathematics.

See figure E.6 (page 90) for actions you can take to motivate yourself, your students, and their parents.

> The following three actions can help you motivate yourself, students, and parents to help girls excel in mathematics.
>
> 1. Write your own letter to your current students to motivate them as mathematics learners in your classroom.
> 2. Come up with a list of 10 of your own affirmations to add to this list.
> 3. Write a letter to the parents of your current students inviting them to partner with you in teaching mathematics.

Figure E.6: Actions teachers can take to motivate themselves, students, and parents.

Reflections

Answer the following five questions independently or in your book study to further your understanding and goals related to teaching girls mathematics.

1. What structures are currently in place at your school to support lessons that promote gender inclusiveness? What are some hindrances at your school to implementing lessons that promote gender inclusiveness? How will gender-inclusive lessons positively impact your school?

2. What project could be started at the classroom level that could benefit your school or community? Why do you want to do this project?

3. What type of project do you feel parents or guardians of your students would like their children to be involved in that will positively impact the community? What role does mathematics play in this project?

4. What will you say the next time you hear someone say that boys are better at mathematics than girls?

5. Who will you partner with in order to bring about change to girls mathematics experiences on a larger scale?

References and Resources

Amelink, C. T. (2012). Female interest in mathematics. In B. Bogue & E. Cady (Eds.), *Apply Research to Practice (ARP) resources*. Accessed at www.engr.psu.edu/AWE/ARPResources.aspx on August 24, 2018.

Ansell, E., & Doerr, H. M. (2000). NAEP findings regarding gender: Achievement, affect, and instructional experiences. In E. A. Silver & P. A. Kenney (Eds.), *Results from the seventh mathematics assessment of the National Assessment of Educational Progress* (pp. 73–106). Reston, VA: National Council of Teachers of Mathematics.

Atkinson, M. (2001, September 25). Eat drink man Mariah. *The Village Voice*. Accessed at www.villagevoice.com/2001/09/25/eat-drink-man-mariah on September 18, 2018.

Azar, B. (2010, July/August). *Math + culture = gender gap?* [Blog post]. Accessed at www.apa.org/monitor/2010/07-08/gender-gap.aspx on August 29, 2018.

Ball, D. L., Hill, H. C., & Bass, H. (2005). Knowing mathematics for teaching: Who knows mathematics well enough to teach third grade, and how can we decide? *American Educator, 29*(1), 14–17.

Beede, D., Julian, T., Langdon, D., McKittrick, G., Khan, B., & Doms, M. (2011). *Women in STEM: A gender gap to innovation* (Issue Brief No. 04–11). Accessed at www.esa.doc.gov/sites/default/files/womeninstemagaptoinnovation8311.pdf on August 24, 2018.

Beilock, S. L., Gunderson, E. A., Ramirez, G., & Levine, S. C. (2010). Female teachers' math anxiety affects girls' math achievement. *Proceedings of the National Academy of Sciences, 107*(5), 1860–1863.

Bhatt, M., Blakley, J., Mohanty, N., & Payne, R. (n.d.). *How media shapes perceptions of science and technology for girls and women*. Accessed at https://learcenter.org/pdf/FEMMediaSTEM.pdf on September 18, 2018.

Bian, L., Leslie, S. J., & Cimpian, A. (2017). Gender stereotypes about intellectual ability emerge early and influence children's interests. *Science, 355*(6323), 389–391.

Bottoms, G., &Schmidt-Davis, J. (2010). *The three essentials: Improving schools requires district vision, district and state support, and principal leadership*. Accessed at www.wallacefoundation.org/knowledge-center/Documents/Three-Essentials-to-Improving-Schools.pdf on August 24, 2018.

Breda, T., Jouini, E., & Napp, C. (2018). Societal inequalities amplify gender gaps in math. *Science, 359*(6381), 1219–1220.

Brown Center on Education Policy. (2013). *The 2013 Brown Center report on American education: How well are American students learning?* Washington, DC: The Brookings Institute.

Campbell, P. F., Nishio, M., Smith, T. M., Clark, L. M., Conant, D. L., Rust, A. H., et al. (2014). The relationship between teachers' mathematical content and pedagogical knowledge, teachers' perceptions, and student achievement. *Journal for Research in Mathematics Education, 45*(4), 419–459.

Carpenter, T. P., Fennema, E., Franke, M. L., Levi, L., & Empson, S. B. (2015). *Children's mathematics: Cognitively guided instruction* (2nd ed.). Portsmouth, NH: Heinemann.

Casey, L. (2017, December 11). *Confidence in math has become a major problem for girls in school*. Accessed at https://globalnews.ca/news/3909143/confidence-in-math-has-become-a-major-problem-for-girls-in-school on September 17, 2018.

Ceci, S. J., & Williams, W. M. (2010). *The mathematics of sex: How biology and society conspire to limit talented women and girls*. New York: Oxford University Press.

Cheema, J. R., & Galluzzo, G. (2013). Analyzing the gender gap in math achievement: Evidence from a large-scale US sample. *Research in Education, 90*(1), 98–112. Accessed at https://doi.org/10.7227/RIE.90.1.7 on December 5, 2018.

Cimpian, J. R., Lubienski, S. T., Timmer, J. D., Makowski, M. B., & Miller, E. K. (2016). Have gender gaps in math closed? Achievement, teacher perceptions, and learning behaviors across two ECLS-K cohorts. *AERA Open, 2*(4), 1–19. Accessed at https://doi.org/10.1177/2332858416673617 on August 24, 2018.

Clarke, D., & Roche, A. (2018). Using contextualized tasks to engage students in meaningful and worthwhile mathematics learning. *Journal of Mathematics Behavior, 51*, 95–108. Accessed at https://doi.org/10.1016/j.jmathb.2017.11.006 on August 24, 2018.

Community Tool Box. (2018). *Section 9: Changing policies in schools*. Accessed at https://ctb.ku.edu/en/table-of-contents/implement/changing-policies/school-policies/main on August 24, 2018.

Corbett, C., & Hill, C. (2015). *Solving the equation: The variables for women's success in engineering and computing*. Washington, DC: American Association of University Women.

Damarin, S., & Erchick, D. B. (2010). Toward clarifying the meanings of gender in mathematics education research. *Journal for Research in Mathematics Education, 41*(4), 310–323.

Dennehy, T. C., & Dasgupta, N. (2017). Female peer mentors early in college increase women's positive academic experiences and retention in engineering. *Proceedings of the National Academy of Sciences, 114*(23), 5964–5969.

Dixon, J. K., Brooks, L. A., & Carli, M. R. (2019). *Making sense of mathematics for teaching the small group*. Bloomington, IN: Solution Tree Press.

Dixon, J. K., Nolan, E. C., & Adams, T. L. (2016). *What does it mean to teach mathematics with focus, coherence, and rigor, and how is it achieved?* Bloomington, IN: Solution Tree Press.

Dixon, J. K., Nolan, E. C., Adams, T. L., Brooks, L. A., & Howse, T. D. (2016). *Making sense of mathematics for teaching grades K–2*. Bloomington, IN: Solution Tree Press.

Dixon, J. K., Nolan, E. C., Adams, T. L., Tobias, J. M., & Barmoha, G. (2016). *Making sense of mathematics for teaching grades 3–5*. Bloomington, IN: Solution Tree Press.

DuFour, R., DuFour, R., Eaker, R., & Many, T. (2006). *Learning by doing: A handbook for Professional Learning Communities at Work*. Bloomington, IN: Solution Tree Press.

DuFour, R., DuFour, R., Eaker, R., Many, T. W., & Mattos, M. (2016). *Learning by doing: A handbook for Professional Learning Communities at Work* (3rd ed.). Bloomington, IN: Solution Tree Press.

Duncan, G. J., Dowsett, C. J., Claessens, A., Magnuson, K., Huston, A. C., Klebanov, P., et al. (2007). School readiness and later achievement (Supplemental). *Developmental Psychology*, *43*(6), 1428–1446. Accessed at https://doi.org/10.1037/0012-1649.43.6.1428.supp on December 5, 2018.

Edelman, M. W. (2015, August 21). It's hard to be what you can't see. *Child Watch*. Accessed at www.childrensdefense.org/child-watch-columns/health/2015/its-hard-to-be-what-you-cant-see/ on December 6, 2018.

Educational Research Center of America. (2016). *STEM classroom to career: Opportunities to close the gap*. Accessed at https://careertech.org/resource/stem-classroom-to-career-report on August 24, 2018.

Ellison, G., & Swanson, A. (2010). The gender gap in secondary school mathematics at high achievement levels: Evidence from the American Mathematics Competitions. *Journal of Economic Perspectives*, *24*(2), 109–128.

Else-Quest, N. M., Hyde, J. S., & Linn, M. C. (2010). Cross-national patterns of gender differences in mathematics: A meta-analysis. *Psychological Bulletin*, *136*(1), 103–127.

Fennema, E., Carpenter, T. P., Jacobs, V. R., Franke, M. L., & Levi, L. W. (1998). A longitudinal study of gender differences in young children's mathematical thinking. *Educational Researcher*, *27*(5), 6–11.

Fennema, E., Peterson, P. L., Carpenter, T. P., & Lubinski, C. A. (1990). Teachers' attributions and beliefs about girls, boys, and mathematics. *Educational Studies in Mathematics, 21*(1), 55–69.

Florida State University. (2017, April 6). Under challenge: Girls' confidence level, not math ability hinders path to science degrees. *ScienceDaily*. Accessed at www.sciencedaily.com/releases/2017/04/170406121532.htm on September 21, 2018.

Fryer, R. G., & Levitt, S. D. (2010). An empirical analysis of the gender gap in mathematics. *American Economic Journal*, *2*(2), 210–240.

Ganley, C. M., & Lubienski, S. T. (2016a, May 9). *Current research on gender differences in math* [Blog post]. Accessed at www.nctm.org/Publications/Teaching-Children-Mathematics/Blog/Current-Research-on-Gender-Differences-in-Math on August 24, 2018.

Ganley, C. M., & Lubienski, S. T. (2016b). Mathematics confidence, interest, and performance: Examining gender patterns and reciprocal relations. *Learning and Individual Differences, 47*, 182–193. Accessed at https://doi.org/10.1016/j.lindif.2016.01.002 on December 5, 2018.

Ganley, C. M., & Lubienski, S. T. (2016c, May 23). *What can we do about gender differences in math?* [Blog post]. Accessed at www.nctm.org/Publications/Teaching-Children-Mathematics/Blog/What-Can-We-Do-about-Gender-Differences-in-Math_ on August 24, 2018.

Gavin, M. K., & Reis, S. M. (2003). Helping teachers to encourage talented girls in mathematics. *Gifted Child Today*, *26*(1), 32–64.

Girl Scout Research Institute. (2012). *Generation STEM: What girls say about science, technology, engineering, and math*. Accessed at www.girlscouts.org/content/dam/girlscouts-gsusa/forms-and-documents/about-girl-scouts/research/generation_stem_full_report.pdf on August 24, 2018.

Gojak, L. M. (2013). *Partnering with parents*. Accessed at www.nctm.org/News-and-Calendar/Messages-from-the-President/Archive/Linda-M_-Gojak/Partnering-with-Parents on August 24, 2018.

Gresalfi, M. S., & Chapman, K. (2017, April). *Recrafting manipulatives: Toward a critical analysis of gender and mathematical practice*. Paper presented at the 9th International Mathematics Education and Society Conference, Volos, Greece.

Gutierrez, R. (2002). Enabling the practice of mathematics teachers in context: Toward a new equity research agenda. *Mathematical Thinking and Learning, 4*(2&3), 145–187.

Halpern, D. F., Aronson, J., Reimer, N., Simpkins, S., Star, J. R., & Wentzel, K. (2007). *Encouraging girls in math and science: IES practice guide* (NCER 2007–2003). Washington, DC: National Center for Education Research, Institute of Education Sciences, U.S. Department of Education. Accessed at https://ies.ed.gov/ncee/wwc/Docs/PracticeGuide/20072003.pdf on August 24, 2018.

Hill, C., Corbett, C., & St. Rose, A. (2010). *Why so few?: Women in science, technology, engineering, and mathematics*. Accessed at www.aauw.org/aauw_check/pdf_download/show_pdf.php?file=why-so-few-research on August 24, 2018.

Hill, H. C., & Lubienski, S. T. (2007). Teachers' mathematical knowledge for teaching and school context: A study of California teachers. *Educational Policy, 21*(5), 747–768.

Hill, J. B. (2016). Questioning techniques: A study of instructional practice. *Peabody Journal of Education, 91*(5), 660–671.

hooks, b. (2003). *Teaching community: A pedagogy of hope*. New York: Routledge.

Howse, T. D. (2013). *A case study exploring the relationship between culturally responsive teaching and a mathematical practice of the common core state standards* (Doctoral dissertation). University of Central Florida, Orlando.

Hyde, J. S., Canning, E. A., Rozek, C. S., Clarke, E., Hulleman, C. S., & Harackiewicz, J. M. (2017). The role of mothers' communication in promoting motivation for math and science course-taking in high school. *Journal of Research on Adolescence, 27*(1), 49–64.

Kiwanuka, H. N., Van Damme, J., Van Den Noortgate, W., Anumendem, D. N., Vanlaar, G., Reynolds, C. et al. (2017). How do student and classroom characteristics affect attitude toward mathematics? A multivariate multilevel analysis. *School Effectiveness and School Improvement, 28*(1), 1–21.

Klass, P. (2017, May 15). No such thing as a math person. *The New York Times*. Accessed at www.nytimes.com/2017/05/15/well/family/trying-to-add-up-girls-and-math.html on September 20, 2018.

Klein, P. S., Adi-Japha, E., & Hakak-Benizri, S. (2010). Mathematical thinking of kindergarten boys and girls: Similar achievement, different contributing processes. *Education Studies in Mathematics, 73*(3), 233–246. Accessed at www.jstor.org/stable/40603169 on December 5, 2018.

Kollmayer, M., Schober, B., & Spiel, C. (2018). Gender stereotypes in education: Development, consequences, and interventions. *European Journal of Developmental Psychology, 15*(4), 361–377.

Krulik, S., & Reys, R. E. (1980). *Problem solving in school mathematics*. Reston, VA: The National Council of Teachers of Mathematics.

Larson, M. R., Fennell, F., Adams, T. L., Dixon, J. K., Kobett, B. M., & Wray, J. A (2012). *Common Core mathematics in a PLC at Work, grades K–2*. T. D. Kanold (Ed.). Bloomington, IN: Solution Tree Press.

Lauzen, M. M., Dozier, D. M., & Horan, N. (2008). Constructing gender stereotypes through social roles in prime-time television. *Journal of Broadcasting and Electronic Media, 52*(2), 200–214.

Lazarides, R., & Watt, H. M. G. (2015). Girls' and boys' perceived mathematics teacher beliefs, classroom learning environments and mathematical career intentions. *Contemporary Educational Psychology, 41*, 51–61. Accessed at https://doi.org/10.1016/j.cedpsych.2014.11.005 on December 5, 2018.

Leonard, J. (2008). *Culturally specific pedagogy in the mathematics classroom: Strategies for teachers and students*. New York: Routledge.

Levine, G. (2013, October 25). *Closing the gender gap: Increasing confidence for teaching mathematics*. Proceedings from the 44th Annual Conference of the Northeastern Educational Research Association, Rocky Hill, Connecticut. Accessed at https://opencommons.uconn.edu/cgi/viewcontent.cgi?article=1006&context=nera_2013 on August 28, 2018.

Leyva, L. A. (2017). Unpacking the male superiority myth and masculinization of mathematics at the intersections: A review of research on gender in mathematics education. *Journal for Research in Mathematics Education, 48*(4), 397–452. Accessed at https://doi:10.5951/jresematheduc.48.4.0397 on December 5, 2018.

Lindberg, S. M., Hyde, J. S., Petersen, J. L., & Linn, M. C. (2010). New trends in gender and mathematics performance: A meta-analysis. *Psychological Bulletin, 136*(6), 1123–1135.

Lloyd, J. E. V., Walsh, J., & Yailagh, M. S. (2005). Sex differences in performance attributions, self-efficacy, and achievement in mathematics: If I'm so smart, why don't I know it? *Canadian Journal of Education/Revue canadienne de l'éducation, 28*(3), 384–408.

Lubienski, S. T., Robinson, J. P., Crane, C. C., & Ganley, C. M., (2013). Girls' and boys' mathematics achievement, affect, and experiences: Findings from ECLS-K. *Journal for Research in Mathematics Education, 44*(4), 634-645. Accessed at https://doi:10.5951/jresematheduc.44.4.0634 on December 5, 2018.

Manna, P., & McGuinn, P. (2013). *Education governance for the twenty-first century: Overcoming the structural barriers to school reform*. Washington, DC: Brookings Institution Press.

Marks, G. N. (2008). Accounting for the gender gaps in student performance in reading and mathematics: Evidence from 31 countries. *Oxford Review of Education, 34*(1), 89–109. Accessed at www.jstor.org/stable/20462373 on December 5, 2018.

Mendick, H. (2005). Mathematical stories: Why do more boys than girls choose to study mathematics at AS-level in England? *British Journal of Sociology of Education, 26*(2), 235–251.

Miller, A. D., Ramirez, E. M., & Murdock, T. B. (2017). The influence of teachers' self-efficacy on perceptions: Perceived teacher competence and respect and student effort and achievement. *Teaching and Teacher Education, 64*, 260–269.

Muijs, D., & Reynolds, D. (2002). Teachers' beliefs and behaviors: What really matters? *The Journal of Classroom Interaction, 37*(2), 3–15. Accessed at www.jstor.org/stable/23870407 on December 5, 2018.

National Center for Education Statistics. (2017). *National Assessment of Educational Progress 2017 mathematics assessments.* Accessed at https://nces.ed.gov/nationsreportcard/mathematics on August 24, 2018.

National Council of Teachers of Mathematics. (2000). *Principles and standards for school mathematics: An overview.* Reston, VA: Author.

National Council of Teachers of Mathematics. (2013). *Formative assessment: A position of the National Council of Teachers of Mathematics.* Reston, VA: Author. Accessed at www.nctm.org/uploadedFiles/Standards_and_Positions/Position_Statements/Formative%20Assessment1.pdf on August 24, 2018.

National Council of Teachers of Mathematics. (2014). *Principles to actions: Ensuring mathematical success for all.* Reston, VA: Author.

National Council of Teachers of Mathematics. (2016, July). *High expectations in mathematics education.* Accessed at www.nctm.org/uploadedFiles/Standards_and_Positions/Position_Statements/High%20Expectations%200816.pdf on December 5, 2018.

National Governors Association Center for Best Practices & Council of Chief State School Officers. (2010). *Standards for mathematical practice.* Washington, DC: Authors.

National Science Board. (2016). *Science and engineering indicators 2016: A broad base of quantitative information on the U.S. and international science and engineering enterprise* (NSB 2016–1). Arlington, VA: National Science Foundation.

National Science Board. (2018). *Early gender gaps in mathematics and teachers' perceptions.* Accessed at www.nsf.gov/statistics/2018/nsb20181/assets/481/early-gender-gaps-in-mathematics-and-teachers-perceptions.pdf on August 22, 2018.

Niederle, M., & Vesterlund, L. (2010). Explaining the gender gap in math test scores: The role of competition. *Journal of Economic Perspectives, 24*(2), 129–144.

Nolan, E. C., Dixon, J. K., Roy, G. J., & Andreasen, J. B. (2016). *Making sense of mathematics for teaching grades 6–8.* Bloomington, IN: Solution Tree Press.

Nolan, E. C., Dixon, J. K., Safi, F., & Haciomeroglu, E. S. (2016). *Making sense of mathematics for teaching high school.* Bloomington, IN: Solution Tree Press.

Noonan, R. (2017). *Women in STEM: 2017 update* (ESA Issue Brief #06–17). Washington, DC: U.S. Department of Commerce, Economics and Statistics Administration.

Nurlu, Ö. (2015). Investigation of teachers' mathematics teaching self-efficacy. *International Electronic Journal of Elementary Education, 8*(1), 21–40.

Organisation for Economic Co-operation and Development. (2013). *The role of families in shaping students' engagement, drive and self-beliefs.* Accessed at www.oecd.org/pisa/keyfindings/PISA2012-Vol3-Chap6.pdf on September 17, 2018.

Organisation for Economic Co-operation and Development. (2016). *Distribution of teachers by age and gender.* Accessed at https://stats.oecd.org/Index.aspx?DataSetCode=EAG_PERS_SHARE_AGE on November 30, 2018.

Overdeck, L. B. (2018). *5 ways to build math into your child's day.* Accessed at www.naeyc.org/our-work/families/5-ways-build-math-your-childs-day on December 5, 2018.

Pearson, N. (n.d.). Different for girls? *International Teacher Magazine.* Accessed at https://consiliumeducation.com/itm/2018/06/29/different-for-girls on August 22, 2018.

Penner, A. M., & Paret, M. (2008). Gender differences in mathematics achievement: Exploring the early grades and the extremes. *Social Science Research, 37*(1), 239–253.

Perception Institute. (2018). *Research: Science and perception.* Accessed at https://perception.org/research/ on December 5, 2018.

Perez-Felkner, L., Nix, S., & Thomas, K. (2017). Gendered pathways: How mathematics ability beliefs shape secondary and postsecondary course and degree field choices. *Frontiers in Psychology, 8,* 386.

Popham, W. J. (2008). *Transformative assessment.* Alexandria, VA: Association for Supervision and Curriculum Development.

Post, G. (2015). *Difficult passage: Gifted girls in middle school.* Accessed at www.davidsongifted.org/Search-Database/entry/A10852 on September 17, 2018.

Reardon, S. F., Fahle, E., Kalogrides, D., Podolsky, A., & Zarate, R. C. (2018). *Gender achievement gaps in U.S. school districts* (CEPA Working Paper No. 18–13). Accessed at https://cepa.stanford.edu/content/gender-achievement-gaps-us-school-districts on August 24, 2018.

Reardon, S. F., Kalogrides, D., Fahle, E. M., Podolsky, A. & Zarate, R. C. (2018). The relationship between test item format and gender achievement gaps on math and ELA tests in fourth and eighth grades. *Educational Researcher, 47*(5), 284–294.

Rellensmann, J., & Schukajlow, S. (2017). Does students' interest in a mathematical problem depend on the problem's connection to reality? An analysis of students' interest and pre-service teachers' judgments of students' interest in problems with and without a connection to reality. *ZDM, 49*(3), 367–378.

Rideout, V. J., Foehr, U. G., & Roberts, D. F. (2010). *Generation M2: Media in the lives of 8- to 18-year-olds* (A Kaiser Family Foundation Study). Accessed at https://files.eric.ed.gov/fulltext/ED527859.pdf on September 18, 2018.

Riegle-Crumb, C., & Humphries, M. (2012). Exploring bias in math teachers' perceptions of students' ability by gender and race/ethnicity. *Gender and Society, 26*(2), 290–322. Accessed at https://doi.org/10.1177/0891243211434614 on December 5, 2018.

Riegle-Crumb, C., & King, B. (2010). Questioning a white male advantage in STEM. *Educational Researcher, 39*(9), 656–664.

Robinson, J. P., & Lubienski, S. T. (2011). The development of gender achievement gaps in mathematics and reading during elementary and middle school: Examining direct cognitive assessments and teacher ratings. *American Educational Research Journal, 48*(2), 268–302.

Robinson-Cimpian, J. P., Lubienski, S. T., Ganley, C. M., & Copur-Gencturk, Y. (2014). Teachers' perceptions of students' mathematical proficiency may exacerbate early gender gaps in achievement. *Developmental Psychology, 50*(4), 1262–1281.

Salikutluk, Z., & Heyne, S. (2017). Do gender roles and norms affect performance in maths?: The impact of adolescents' and their peers' gender conceptions on maths grades. *European Sociological Review, 33*(3), 368–381.

Sawchuk, S. (2018, June). *A bold effort to end algebra tracking shows promise.* Accessed at www.edweek.org/ew/articles/2018/06/13/a-bold-effort-to-de-track-algebra-shows.html on December 11, 2018.

Schaffhauser, D. (2018, September 13). *Gender gap among math high achievers evident by grade 9, just gets wider.* Accessed at https://thejournal.com/articles/2018/09/13/gender-gap-among-math-high-achievers-evident-by-grade-9.aspx on December 5, 2018.

Schmidt, W. H., Burroughs, N. A., Cogan, L. S., & Houang, R. T. (2017). The role of subject-matter content in teacher preparation: An international perspective for mathematics. *Journal of Curriculum Studies, 49*(2), 111–131.

Schwartz, S. (2018, July 19). *To up students' math ability, try working on their teachers' growth mindset* [Blog post]. Accessed at https://blogs.edweek.org/teachers/teaching_now/2018/07/up_students_math_ability_teachers_growth_mindset.html?print=1 on August 24, 2018.

Skemp, R. R. (1978). Relational understanding and instrumental understanding. *Arithmetic Teacher, 26*(3), 9–15.

Small, M. (2009). *Good questions: Great ways to differentiate mathematics instruction.* New York: Teachers College Press.

Smith, M. S., & Stein, M. K. (2011). *5 practices for orchestrating productive mathematics discussions.* Reston, VA: National Council of Teachers of Mathematics.

Smith, S. S. (2009). *Using manipulatives: Learn how to effectively use fraction strips, spinners, counters, and more.* Accessed at www.teachervision.fen.com/pro-dev/teaching-methods/48934.html on August 24, 2018.

Spencer, S. J., Steele, C. M., & Quinn, D. M. (1999). Stereotype threat and women's math performance. *Journal of Experimental Social Psychology, 35*(1), 4–28.

Soni, A., & Kumari, S. (2017). The role of parental math anxiety and math attitude in their children's math achievement. *International Journal of Science and Mathematics Education, 15*(2), 331–347.

Steinke, J. (2017). Adolescent girls' STEM identity formation and media images of STEM professionals: Considering the influence of contextual cues. *Frontiers in Psychology, 8*, 716.

Steinke, J., & Tavarez, P. M. P. (2016). *Cultural representations of gender and science: portrayals of female STEM professionals in popular films 2002–2015.* Paper Presented at the Association for Education in Journalism and Mass Communication, San Francisco, CA.

Stoehr, K. J. (2017). Mathematics anxiety: One size does not fit all. *Journal of Teacher Education, 68*(1), 69–84.

Stoet, G., & Geary, D. C. (2018). The gender-equality paradox in science, technology, engineering, and mathematics education. *Psychological Science, 29*(4), 581–593.

Tiedemann, J. (2000). Parents' gender stereotypes and teachers' beliefs as predictors of children's concept of their mathematical ability in elementary school. *Journal of Educational Psychology, 92*(1), 144–151.

Tzuriel, D., & Egozi, G. (2010). Gender differences in spatial ability of young children: The effects of training and processing strategies. *Child Development, 81*(5), 1417–1430.

Uribe-Flórez, L. J., & Wilkins, J. L. M. (2017). Manipulative use and elementary schools students' mathematics learning. *International Journal Science and Math Education, 15*(8), 1541–1557.

Villalobos, A. (2009). The importance of breaking set: Socialized cognitive strategies and the gender discrepancy in mathematics. *Theory and Research in Education, 7*(1), 27–45.

Wallace, K. (2016, October 12). *The "boys are better at math" mindset creates gender gap in sciences*. Accessed at www.cnn.com/2016/10/12/health/female-scientists-engineers-math-gender-gap/index.html on December 5, 2018.

Walshaw, M., & Anthony, G. (2008). The teacher's role in classroom discourse: A review of recent research into mathematics classrooms. *Review of Educational Research, 78*(3), 516–551.

Wang, M. T., & Degol, J. L. (2017). Gender gap in science, technology, engineering, and mathematics (STEM): Current knowledge, implications for practice, policy, and future directions. *Educational Psychology Review, 29*(1), 119–140.

Ward, L. M., & Aubrey, J. S. (2017). *Watching gender: How stereotypes in movies and on TV impact kids' development*. San Francisco: Common Sense. Accessed at www.k12blueprint.com/sites/default/files/2017_commonsense_watchinggender_executivesummary_0620_1.pdf on September 18, 2018.

Webb, N. M., Franke, M. L., Ing, M., Turrou, A. C., Johnson, N. C., & Zimmerman, J. (2017). Teacher practices that promote productive dialogue and learning in mathematics classroom. *International Journal of Educational Research*.

Wiest, L. R. (2014). *Strategies for parents to support daughters in STEM*. Accessed at www.unr.edu/education/centers-student-resources/initiatives/girls-math-camp/resources/parents/tips on December 5, 2018.

Yoder, N. (2014). *Teaching the whole child: Instructional practices that support social-emotional learning in three teacher evaluation frameworks* (rev. ed.) (Research-to-Practice Brief). Washington DC: Center on Great Teachers and Leaders at American Institutes for Research. Accessed at https://gtlcenter.org/sites/default/files/TeachingtheWholeChild.pdf on August 24, 2018.

Index

A

ability differences, 50–51. *See also* achievement gap
achievement
    data, use of, 25, 27
    impact of early, 81
    recommendations for instruction supporting, 24–25
achievement gap, 2–3, 7–18, 9–10
    confidence levels and, 15
    evidence against, 14–15
    evidence pointing to, 10–14
    in mathematics scores, 11–12
    race and, 12
    reflection on, 8–9
    research on, interpretations of, 26
    teachers' perceptions of, 15–16, 73–74
action strategies, 79
Adams, T. L., 1, 50, 61, 71, 72
affirmations, 86–88
algorithmic strategies, 13
Answell, E., 11–12
Anthony, G., 37
anxiety, 15, 16
arguments, constructing, 35, 44
assessment. *See* formative assessment
attitudes, 25, 26, 27
authenticity, 69, 85, 87–88
Azar, B., 26

B

background knowledge, 50–51
Ball, D., 79
Bass, H., 79
Beede, D., 11
Beilock, S., 13
beliefs
    about equity, 69
    defining, 20–21
    educator definitions of, 19–20
    perception and, 19–22
    priorities and, 73–76
bias. *See also* stereotypes
    defining, 21
    educator definitions of, 19–20
    perception and, 19–22
    in task context, 46–48
    work-life, 76
Brooks, L., 71, 72
Brown Center on Education Policy, 77–78

C

Carpenter, T., 13
Casey, L., 12
Cimpian, J., 73, 81
Clarke, D., 46
classrooms
    fostering discourse in, 37–39
    perceptions in, 22–25
    planning positive practices in, 52–60
collaboration
    with families, 64
    to provide opportunities for girls in STEM, 77
collaborative culture, 61
collective responsibility, 61
communication
    about achievement data, 26
    about progress, 72–73
    of expectations, 50–52
    with families, 30
    of lack of understanding, 23
Community Tool Box, 77–79
confidence, 2
    achievement and, 15
    actions to build, 31
    modeling mathematical power and, 48–49
    parents/guardians and, 30
    teacher content knowledge and, 81
    of teachers, 13, 23, 26

teaching emphasis on, 60
consensus, 40
content knowledge, 62, 79–81
context, 46–48
Coper-Gencturk, Y., 16
Corbett, C., 22
Council of Chief State School Officers (CCSSO), 35, 37–38
critical thinking, 37, 39, 41–43, 45–48, 49, 80
cultural differences, 30, 60
    equity and, 68–73
    including in tasks, 72

D

differential ratings, 16
differentiated instruction, 36
    equity and, 68–73
disabilities, students with, 68–69
disagreements, 51
discourse, classroom, 23–24, 34
    fostering, 37–39
    questioning and, 41
    student selection in, 49
discourse cards, 52–53
districts
    perceptions in, 25–27
    possibilities in, 60–62
diversity, honoring, 34, 36–37
    interest surveys and, 55, 57–58
    positive practices for, 52–53
Dixon, J., 1, 50, 61, 71, 72
Doerr, H., 11–12
Do Now, 5

E

Edelman, M., 10–11
Educational Research Center of America, 25
Education Quality and Accountability Office, 12
educators
    confidence of, 13
    content knowledge of, 62
    conveying expectations of, 35, 49–52
    definitions of perceptions, beliefs, bias, stereotypes, 19–20
    expectations of, 23–24
    on girls in mathematics, 7–8
    honoring diverse ways of doing mathematics, 34, 36–37
    knowledge of, 79–81
    mindsets of, 15–16
    motivation for and by, 89–90
    open letters to, 85, 86
    planning for positive practices for girls, 51–64
    priorities for, 67–83
    professional development for, 62
    recommendations for mathematics instruction, 24–25
    school/district-level, 25–27
    in school-home connection, 27–30
    self-efficacy of, 81
    in student success, 34
    support for policies from, 78–79
    women as, 13, 26
Elysee, D., 28
empowerment, 67–68
encouragement, 59, 85–90
    affirmations, 85, 87–88
    resources for, 85–89
"Encouraging Girls in Math and Science" (Halpern et al.), 24, 25
engagement
    equity and, 70
    finding evidence of, 37, 39, 40, 43, 44–45, 47, 49, 51
    formative assessment and, 44
    nurturing, 58
    with problem solving, 13
    in tasks, 71–73
    teacher actions to encourage, 36, 38, 40, 42, 44, 47, 48–49, 50–51
environment, safe and inclusive, 23–24
    authenticity and, 69
    challenging, 80
    nurturing, 58–59
    student-teacher relationships and, 55
equality, equity vs., 68
equity, 67, 68–73
Erickson, K., 33
evidence. See also TQE process
    on the achievement gap, 10–15
    of engagement, 37, 39, 40, 43, 44–45, 47, 49, 51
    equity and, 71–73
    of learning, 72
expectations, 23–24
    conveying, 35, 49–52

cultural differences and, 60
    self-reflection chart for, 58, 59

F

Fahle, E., 12
feedback, 24, 59
Fennema, E., 13
focus on learning, 61
formative assessment, 35, 43–46, 55, 56
Franke, M., 13

G

Ganley, C., 2, 16, 25, 26
Gavin, M., 49–50
gender differences, 1
    in achievement, 2–3, 7–18
    in anxiety, 15, 16
    instructional approach and, 2
    in mathematics achievement, 2–3, 11–12
    messaging on, 25–27
    in problem-solving approaches, 13
    reflection on, 8–9
    research on, interpretations of, 26
    self-concept in mathematics and, 12
    in spatial skills, 14
    in STEM representation, 10–11, 33
    stereotypes on, 15
    systemic messaging on, 25–27
    task context and, 46–48
    teacher expectations on, 49–52
gender identification, 1
gender inclusiveness, 70–73
Girl Scout Research Institute, 76, 85
Girls Who Code, 33, 77
goals
    for communicating data, 26
Good Questions: Great Ways to Differentiate Mathematics Instruction (Small), 36
governance, 77–78
grants, 77
guest speakers, 77
Gunderson, E., 13

H

hands-on learning, 35, 39–41
    mathematics toolbox for, 53–54
Hill, C., 22

Hill, H., 79, 80
Hill, J., 41
homework, 63–64
honor societies, 77
Howse, T., 71, 72

I

inclusiveness, 2, 23–24, 70–73
intelligence, 15
interest in mathematics, 2
    fostering, 24
    interest in STEM fields and, 76
    maintaining girls', 9–10
    parents/guardians and, 29
    student interest surveys and, 55, 57–58
interest surveys, 55, 57–58

J

Jacobs, V., 13
Johnson, K., 7

K

Kalogrides, D., 12
King, B., 10
knowledge
    building on, 72
    content, of teachers, 41, 50–51
    of parents, 63–64
    student's background, 50–51, 73–74
    of teachers, 79–81

L

learning
    belief statements about, 73–74
    evidence of, 72
    focus on, 61
    from "no," 85–86
    progression of, 80
letters to students, 85, 86
Levi, L., 13
Levine, S., 13
Lloyd, J., 14
Lubienski, S., 2, 16, 25, 26, 73, 80

M

Maher, J., 62

making sense, 35
Making Sense of Mathematics for Teaching books, 1, 2
Making Sense of Mathematics for Teaching the Small Group (Dixon, Brooks, Carli), 1
Makowski, M., 73
manipulatives, 38, 39–41
mathematics toolbox, 53–54
mentors, 77
Miller, E., 73
mindsets, 15–16, 25
misconceptions, addressing, 38
modeling mathematical power, 35, 46, 48–49, 58
Montessori, M., 33
motivation, 29, 90

N

NAEP. *See* National Assessment of Educational Progress (NAEP)
National Assessment of Educational Progress (NAEP), 11–12
National Center for Education Statistics (NCES), 11, 14–15
National Council for Teachers of Mathematics (NCTM), 14, 43, 50, 68, 73–74
National Governors Association (NGA), 35, 37–38
National Science Board, 25
NCES. *See* National Center for Education Statistics (NCES)
NCTM. *See* National Council for Teachers of Mathematics (NCTM)
newsletters, 62–63
Niederle, M., 26
Nolan, E., 1, 50, 61, 71, 72
"no," learning from, 85–86
Noonan, R., 10
norms
    for classrooms, 23–24
    of listening, 38
    for parents/guardians, 28–29
    for schools and districts, 25–27

O

open letters, 85–86
opportunities, 67, 76–79
    exposing girls to, 9–10
oral language development, 68
outsiders, feeling like, 67

P

parents and guardians
    actions to improve perspectives of, 16–17
    building relationships with, 64
    influence of, 62
    perceptions of, 27–30
    possibilities with, 62–64
    school-home connection with, 27–30
    support for policies and, 78
pedagogical features, 5
pedagogical tools
    discourse cards, 52–53
    mathematics toolbox, 53–54
Perception Institute, 22
perceptions, 3, 19–32
    belief, bias, stereotypes, and, 19–22
    building self-confidence and, 30, 31
    in the classroom, 22–25
    defining, 19–20
    educator definitions of, 19–20
    importance of, 4
    norms for, 23–24
    of parents/guardians, 27–30
    school and district, 25–27
    self-reflection quiz on, 74–76
    unearthing, about girls in mathematics, 22–30
perseverance, 48–49, 85–87
pick-a-word student activity, 89
planning, 35, 39–41
    within PLCs, 61
    for positive teaching practices, 51–64
    for school-home relationships, 64
play button symbol, 4–5
PLC. *See* professional learning community (PLC) process
Podolsky, A., 12
policies, 77–78
Popham, W., 43
possibilities, 3, 4, 33–66
    actions for considering, 64–65
    context and, 46–48
    conveying teacher expectations, 49–52
    formative assessment, 35, 43–46
    fostering classroom discourse, 37–39
    honoring diverse ways of doing mathematics, 34, 36–37
    modeling mathematical power, 35, 46, 48–49

for parents' support, 29
planning for hands-on learning, 39–41
planning for positive practices for girls, 51–52, 52–64
reflection on, 65
standards and, 35
using questioning to boost understanding, 41–43
power, modeling mathematical, 35, 46, 48–49, 58
praise, 15
precision, 35
Principles to Action (NCTM), 73–74
priorities, 3, 4, 67–83
actions for implementing, 81–82
equity, 67, 68–73
opportunity, 67, 76–79
reflections on, 82
teacher beliefs, 67, 73–76
problem solving, 25
alternative approaches in, 38
equity and, 70
gender differences in approaches to, 13
instruction supporting strategies for, 26
modeling mathematical power and, 48–49
sharing strategies for, 52
validating different strategies for, 51
Problem Solving in School Mathematics (Krulik & Reys), 13
professional development, 62
professional learning community (PLC) process, 60–62
progression of learning, 80
prompts, 49

Q

questioning, 35, 41–43. *See also* TQE process
advancing, 42, 43, 54–55
assessing, 42, 43, 44, 54
critical thinking and, 41–43
discourse cards and, 53
facilitative, 41, 47
follow-up, 38, 39
gender bias in, 59
honoring diverse ways of doing mathematics and, 36
inclusiveness in, 71–73
purpose in, 72
Quinn, D., 76

R

race and ethnicity, achievement and, 12
Ramirez, G., 13
Reardon, S., 12
reasoning, 35, 51
formative assessment and, 44
questioning to elicit, 41–43
regularity in repeated, 35
reflection, 2, 5
on anxiety, 15, 16
on communication of progress, 72–73
on equity, 70
on expectations, 58, 59
on formative assessment, 46
on the gender achievement gap, 17
on girls as learners of mathematics, 8–9
on girls' experiences with mathematics, 9
on girls' perception of mathematics, 29
on girls you are teaching, 55, 56
on goals, 90
on perceptions of girls' ability to learn mathematics, 20, 31
on possibilities, 65
student success and, 34
on teaching practices, tools, and programs, 34
Reis, S., 49–50
research
further reading on, 5
resources for teachers, 80, 85–89
results orientation, 61
Riegle-Crumb, C., 10
risk taking, 51–52
Robinson-Cimpian, J., 16
Roche, A., 46

S

San Francisco Unified School District, 77–78
Saujani, R., 33
schools
perceptions in, 25–27
possibilities in, 60–62
Schwartz, S., 13
self-concept, 12
self-efficacy, 81
sense making, 51
skills, as learnable, 24
Small, M., 36

small groups, 1
Smithies, L., 28
solutions, 35
    providing explanations/justifications for, 23
    sharing, 38
    understanding classmates', 23
spatial skills, 14, 24–25
    instruction supporting, 26
    teaching emphasis on, 60
Spencer, S., 76
sponsorships, 77
Standards for Mathematical Practice, 35
Steele, C., 76
STEM
    barriers to girls pursuing, 76
    exposing girls to women in, 24, 33
    girls' perceptions of women in, 85
    increasing opportunities in, 76–79
    representation of women in, 10–11
    spatial skills and, 14
stereotypes, 15, 76
    activities based on, 24
    classroom, 22
    defining, 21–22
    educator definitions of, 19–20
    perception and, 19–22
St. Rose, A., 22
structure, looking for and making use of, 35
struggle
    learning from, 85–86
    role of in education, 15
    teacher response to, 50
success, 59

T

Take Action, 5
tasks. *See also* TQE process
    context for, bias in, 46–48
    equity in, 71–73
    inclusiveness in, 70–73
    interest surveys and, 57–58
teachers. *See* educators
    content understanding/knowledge of, 41, 62
teacher-student relationships, 16
teaching, belief statements about, 73–74
team-building activities, 60
time, 51

Timmer, J., 73
tools. *See also* questioning
    mathematics toolbox, 53–54
    for modeling mathematical power, 57, 58
    teaching use of, 35, 40
TQE process, 71–73
tracking policies, 77–79

U

understanding
    communicating lack of, 23
    teaching emphasis on, 60
    using questioning to boost, 35, 54
Uribe-Flórez, L., 39–40
U.S. Department of Education, 24–25

V

Vesterlund, L., 26
videos
    on area problems, 50–51
    on defining and classifying squares and rectangles, 38–39
    on finding the volume of rectangular prisms, 36–37
    on fractions, 48–49
    of interpreting remainder, 46–48
    on multidigit addition, 44–45
    on place value, 40–41
    on word problems, 42–43
Villalobos, A., 13
visualization, 39–40, 60

W

Walsh, J., 14
Walshaw, M., 37
Webb, N., 37
Wiest, L., 67
Wilkins, J., 39–40
"Women in STEM: 2017 Update" (Noonan), 10
work-life balance bias, 76

Y

Yailagh, M., 14

Z

Zárate, R., 12

Making Sense of Mathematics for Teaching Grades K–2
Juli K. Dixon, Edward C. Nolan, Thomasenia Lott Adams, Lisa A. Brooks, and Tashana D. Howse
Develop a deep understanding of mathematics. With this user-friendly resource, grades K–2 teachers will explore strategies and techniques to effectively learn and teach significant mathematics concepts and provide all students with the precise, accurate information they need to achieve academic success.
BKF695

Making Sense of Mathematics for Teaching Grades 3–5
Juli K. Dixon, Edward C. Nolan, Thomasenia Lott Adams, Jennifer M. Tobias, and Guy Barmoha
Develop a deep understanding of mathematics. With this user-friendly resource, grades 3–5 teachers will explore strategies and techniques to effectively learn and teach significant mathematics concepts and provide all students with the precise, accurate information they need to achieve academic success.
BKF696

Making Sense of Mathematics for Teaching Grades 6–8
Edward C. Nolan, Juli K. Dixon, George J. Roy, and Janet B. Andreasen
Develop a deep understanding of mathematics. With this user-friendly resource, grades 6–8 teachers will explore strategies and techniques to effectively learn and teach significant mathematics concepts and provide all students with the precise, accurate information they need to achieve academic success.
BKF697

Making Sense of Mathematics for Teaching the Small Group
Juli K. Dixon, Lisa A. Brooks, and Melissa R. Carli
Make sense of effective characteristics of K–5 small-group instruction in mathematics. Connect new understandings to classroom practice through the use of authentic classroom video of pulled small groups in action. Use the TQE (Tasks, Questions, Evidence) process to plan time effectively for small-group instruction.
BKF832

Making Sense of Mathematics for Teaching to Inform Instructional Quality
Melissa D. Boston, Amber G. Candela, and Juli K. Dixon
Discover a clear path for improving mathematics instruction at any grade level. Designed for individuals or collaborative teams, this practical resource delivers a toolkit of rubrics, strategies, and videos that readers can rely on to help guide reflections, conversations, feedback, and planning.
BKF834

Solution Tree | Press

Visit SolutionTree.com or call 800.733.6786 to order.

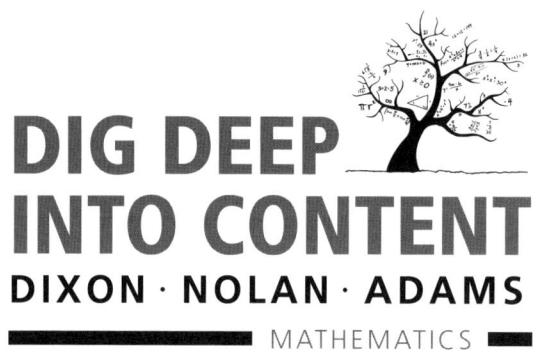

DIG DEEP INTO CONTENT
DIXON · NOLAN · ADAMS
MATHEMATICS

Bring Dixon Nolan Adams Mathematics experts to your school

Juli K. Dixon

Edward C. Nolan

Thomasenia Lott Adams

Janet B. Andreasen
Guy Barmoha
Lisa Brooks
Kristopher Childs
Craig Cullen
Brian Dean

Lakesia L. Dupree
Jennifer Eli
Erhan Selcuk Haciomeroglu
Tashana Howse
Stephanie Luke
Amanda Miller

Samantha Neff
George J. Roy
Farshid Safi
Jennifer Tobias
Taylar Wenzel

Our Services

1. Big-Picture Shifts in Content and Instruction

Introduce content-based strategies to transform teaching and advance learning.

2. Content Institutes

Build the capacity of teachers on important concepts and learning progressions for grades K–2, 3–5, 6–8, and 9–12 based upon the *Making Sense of Mathematics for Teaching* series.

3. Implementation Workshops

Support teachers to apply new strategies gained from Service 2 into instruction using the ten high-leverage team actions from the *Beyond the Common Core* series.

4. On-Site Support

Discover how to unpack learning progressions within and across teacher teams; focus teacher observations and evaluations on moving mathematics instruction forward; and support implementation of a focused, coherent, and rigorous curriculum.

Evidence of Effectiveness

Pasco County School District | Land O' Lakes, FL

Demographics
- 4,937 Teachers
- 68,904 Students
- 52% Free and reduced lunch

Discovery Education Benchmark Assessments

Grade	EOY 2014 % DE	EOY 2015 % DE
2	49%	66%
3	59%	72%
4	63%	70%
5	62%	75%

> "The River Ridge High School Geometry PLC went from ninth out of fourteen high schools in terms of Geometry EOC proficiency in 2013–2014 to first out of fourteen high schools in Pasco County, Florida, for the 2014–2015 school year."
>
> —Katia Clouse, Geometry PLC leader, River Ridge High School, New Port Richey, Florida

Contact your local representative
888.409.1682